Fundamentals of Materials Science and Engineering

Fundamentals of Materials Science and Engineering

Diane Jeffers

Willford Press
www.willfordpress.com

Published by Willford Press,
118-35 Queens Blvd., Suite 400,
Forest Hills, NY 11375, USA

ISBN: 978-1-64728-343-8

Cataloging-in-Publication Data

Fundamentals of materials science and engineering / Diane Jeffers.
p. cm.
Includes bibliographical references and index.
ISBN 978-1-64728-343-8
1. Materials. 2. Materials science. I. Jeffers, Diane.

TA403 .F86 2022
620.11--dc23

For information on all Willford Press publications
visit our website at www.willfordpress.com

Willford Press

Contents

Preface

Materials science is an interdisciplinary field which is focused on designing and discovering new materials. The study of the structure of materials and their properties forms the basis of materials science. A few of the aspects studied under this discipline are crystallography, bonding, thermodynamics and kinetics. There are several research areas which are covered under this discipline such as nanomaterials, biomaterials and semiconductors. Materials science also contributes in the extraction of materials and their conversion into a useful form. The examination of metal alloys is also an important part of materials science. The topics included in this book are of utmost significance and bound to provide incredible insights to readers. It is appropriate for students seeking detailed information in this area as well as for experts. For all those who are interested in materials science, this textbook can prove to be an essential guide.

Given below is the chapter wise description of the book:

Chapter 1- A Chemical substance or a mixture of substances which make up an object is known as a material. The branch of engineering and science which deals with the discovery and design of new materials, especially solids, is referred to as materials science. This chapter briefly examines the key concepts of materials and materials science to provide an introduction to the subject.

Chapter 2- Materials are generally classified on the basis of their properties into metals, nonmetals, metalloids, alloys, composites, polymers and ceramics. Metals are further divided into ferrous metals and non-ferrous metals. The topics elaborated in this chapter will help in gaining a better perspective about these types of materials.

Chapter 3- The properties of materials are divided into several categories in order to make their study more efficient. Some of these are mechanical, chemical, electrical, thermal, optical and magnetic properties. This chapter closely examines these properties of materials to provide an extensive understanding of the subject.

Chapter 4- Bonding in materials refers to the lasting attraction among atoms, molecules or ions which facilitates the creation of chemical compounds. They are broadly divided into primary bonds and secondary bonds. The topics elaborated in this chapter will help in gaining a better perspective about the different types of bonds such as ionic, covalent and hydrogen bonds.

Chapter 5- Material science is an interdisciplinary field which has numerous sub-fields. A few of them are ceramic engineering, metallurgy, polymer science and engineering, and nanotechnology. This chapter has been carefully written to provide an easy understanding of these sub-fields of materials science.

Indeed, my job was extremely crucial and challenging as I had to ensure that every chapter is informative and structured in a student-friendly manner. I am thankful for the support provided by my family and colleagues during the completion of this book.

Diane Jeffers

1

Introduction to Materials Science and Engineering

A Chemical substance or a mixture of substances which make up an object is known as a material. The branch of engineering and science which deals with the discovery and design of new materials, especially solids, is referred to as materials science. This chapter briefly examines the key concepts of materials and materials science to provide an introduction to the subject.

Materials

Materials are the matter or substance that objects are made from. Engineering materials refers to the group of materials that are used in the construction of manmade structures and components. The primary function of an engineering material is to withstand applied loading without breaking and without exhibiting excessive deflection. The major classifications of engineering materials include metals, polymers, ceramics, and composites.

Materials for Energy

An industrially advanced society uses energy and materials in large amounts. Transportation, heating and cooling, industrial processes, communications—in fact, all the physical characteristics of modern life—depend on the flow and transformation of energy and materials through the techno-economic system. These two flows are inseparably intertwined and form the lifeblood of industrial society. The relationship of materials science to energy usage is pervasive and complex. At every stage of energy production, distribution, conversion, and utilization, materials play an essential role, and often special materials properties are needed. Remarkable growth in the understanding of the properties and structures of materials enables new materials, as well as improvements of old ones, to be developed on a scientific basis, thereby contributing to greater efficiency and lower costs.

Classification of Energy-related Materials

Energy materials can be classified in a variety of ways. For example, they can be divided into materials that are passive or active. Those in the passive group do not take part in the actual energy-conversion process but act as containers, tools, or structures such as reactor vessels, pipelines, turbine blades, or oil drills. Active materials are those that take part directly in energy conversion—such as solar cells, batteries, catalysts, and superconducting magnets.

Another way of classifying energy materials is by their use in conventional, advanced, and possible future energy systems. In conventional energy systems such as fossil fuels, hydroelectric generation, and nuclear reactors, the materials problems are well understood and are usually associated with structural mechanical properties or long-standing chemical effects such as corrosion. Advanced energy systems are in the development stage and are in actual use in limited markets. These include oil from shale and tar sands, coal gasification and liquefaction, photovoltaic, geothermal energy, and wind power. Possible future energy systems are not yet commercially deployed to any significant extent and require much more research before they can be used. These include hydrogen fuel and fast-breeder reactors, biomass conversion, and superconducting magnets for storing electricity.

Classifying energy materials as passive or active or in relation to conventional, advanced, or future energy systems is useful because it provides a picture of the nature and degree of urgency of the associated materials requirements. But the most illuminating framework for understanding the relation of energy to materials is in the materials properties that are essential for various energy applications. Because of its breadth and variety, such a framework is best shown by examples. In oil refining, for example, reaction vessels must have certain mechanical and thermal properties, but catalysis is the critical process.

Materials for Ground Transportation

The global effort to improve the efficiency of ground transportation vehicles, such as automobiles, buses, trucks, and trains, and thereby reduce the massive amounts of pollutants they emit, provides an excellent context within which to illustrate how materials science functions to develop new or better materials in response to critical human needs. For the automobile industry in particular, the story is a fascinating one in which the desire for lower vehicle weight, reduced emissions, and improved fuel economy has led to intense competition among aluminum, plastics, and steel companies for shares in the enormous markets involved (40 million to 50 million cars and trucks per year worldwide). In this battle, materials scientists have a key role to play because the success of their efforts to develop improved materials will determine the shape and viability of future automobiles.

Just how seriously suppliers to the industry view the need either to protect or to increase their share of these enormous markets is demonstrated by their establishing of special programs, consortia, or centres that are specifically designed to develop better alloys, plastics, or ceramics for automotive applications. For example, in the United States a program at the Aluminum Company of America (Alcoa) called the aluminum intensive vehicle (AIV), and a similar one at Reynolds Metals, were established to develop materials and processes for making automobile "space frames" consisting of aluminum-alloy rods and die-cast connectors joined by welding and adhesive bonding. Not to be outdone, another aluminum company, Alcan Aluminium Limited of Canada, in a program entitled aluminum structured vehicle technology (ASVT), began to investigate the construction of automobile unibodies from adhesively bonded aluminum sheet. The plastics industry, of course, has a powerful interest in replacing as many metal automobile components as possible, and in order to help bring this about a centre called D&S Plastics International was formed in the Detroit, Mich., area of the United States by three corporations. The specific aim of this centre was to develop materials and a process suitable for forming several connected panels or components (*e.g.*, body panels and bumper fascias) simultaneously out of different types of plastics. The centrepiece of the operation was a 4,000-ton co-injection press that could lead to cost reductions as great as 50 percent and thereby make the use of plastics for automotive applications more attractive.

In programs such as these, and in many more carried out by vendors and within the automobile companies themselves, materials scientists with specialized training in advanced metals, plastics, and ceramics have been leading a revolution in the automotive industry. The following sections describe specific needs that have been identified for improving the performance of automobiles and other ground-transportation vehicles, as well as approaches that materials scientists have taken in response to those needs.

Metals

Aluminum

Since aluminum has about one-third the density of steel, its substitution for steel in automobiles would seem to be a sensible approach to reducing weight and thereby increasing fuel economy and reducing harmful emissions. Such substitutions cannot be made, however, without due consideration of significant differences in other properties of the two materials. This is one important facet of the materials scientist's job—to help evaluate the suitability of a material for a given application based on how its properties balance against load and performance requirements specified by the design engineer. In this case (aluminum versus steel), it is instructive to consider the materials scientist's approach to evaluating the use of aluminum in automotive panels—such components as doors, hoods, trunk decks, and roofs that can make up more than 60 percent of a vehicle's weight.

Two primary properties of any metal are (1) its yield strength, defined as its ability to resist permanent deformation (such as a fender dent), and (2) its elastic modulus, defined as its ability to resist elastic or springy deflection like a drum head. By alloying, aluminum can be made to have a yield strength equal to a moderately strong steel and therefore to exhibit similar resistance to denting in an automobile panel. On the other hand, alloying does not normally affect the elastic modulus of metals significantly, so that automotive door panels or hoods made from aluminum alloys, all of which have approximately one-third the modulus of steel, would be floppy and suffer large deflections when buffeted by the wind, for example. From this point of view, aluminum would appear to be a marginal choice for body panels.

One might attempt to overcome this deficiency by increasing the thickness of the aluminum sheet stock to three times the thickness of the steel it is intended to replace. This, however, would simply increase the weight to roughly that of an equivalent steel structure and thus defeat the purpose of the exercise. Fortunately, as was elegantly demonstrated in 1980 by two British materials scientists, Michael Ashby and David Jones, when proper account is taken of the way an actual door panel deflects, constrained as it is by the door edges, it is possible to use aluminum sheet only slightly thicker than the steel it would replace and still achieve equivalent performance. The net result would be a weight savings of almost two-thirds by the substitution of aluminum for steel on such body components. This suggests that understanding the interrelationship between materials properties and structural design is an important factor in the successful application of materials science.

Another important activity of the materials scientist is that of alloy development, which in some cases involves designing alloys for very specific applications. For example, in Alcoa's AIV effort, materials scientists and engineers developed a special casting alloy for use as cast aluminum nodes (connecters) in their space frame design. Ordinarily, metal castings exhibit very little toughness, or ductility, and they are therefore prone to brittle fracture followed by catastrophic failure. Since the integrity of an automobile would be limited by having relatively brittle body components, a proprietary casting alloy and processing procedure were developed that provide a material of much greater ductility than is normally available in a casting alloy.

Many other advances in aluminum technology, brought about by materials scientists and design engineers, have led to a greater acceptance of aluminum in automobiles, trucks, buses, and even light rail vehicles. Among these are alloys for air-conditioner components that are designed to be chemically compatible with environmentally safer refrigerants and to withstand the higher pressures required by them. Also, alloys have been developed that combine good formability and corrosion resistance with the ability to achieve maximum strength without heat treating; these alloys develop their strength during the forming operation. As a consequence, the list of vehicles that contain significant quantities of aluminum substituted for steel has steadily grown. A milestone was reached in 1992 with a limited-edition Jaguar sports car that was virtually all aluminum, including the engine, adhesively bonded chassis, and skin. Somewhat less expensive and in full production were Honda's Acura NSX, containing more than 400 kilograms (900 pounds) of aluminum compared with about 70 kilograms for the average automobile, and General Motors' Saturn, with an aluminum engine block and cylinder heads. These vehicles and others took their place alongside the British Land Rover, which was built with all-aluminum body panels beginning in 1948—a choice dictated by a shortage of steel during World War II and continued by the manufacturer ever since.

Steel

While the goal of the aluminum and plastics industries is to achieve vehicle weight reductions by substituting their products for steel components, the goal of the steel industry is to counter such inroads with such innovative developments as high-strength, but inexpensive, "microalloyed" steels that achieve weight savings by thickness reductions. In addition, alloys have been developed that can be tempered (strengthened) in paint-baking ovens rather than in separate and expensive heat-treatment furnaces normally required for conventional steels.

The micro alloyed steels, also known as high-strength low-alloy (HSLA) steels, are intermediate in composition between carbon steels, whose properties are controlled mainly by the amount of carbon they contain (usually less than 1 percent), and alloy steels, which derive their strength, toughness, and corrosion resistance primarily from other elements, including silicon, nickel, and manganese, added in somewhat larger amounts. Developed in the l960s and resurrected in the late 1970s to satisfy the need for weight savings through greater strength, the HSLA steels tend to be low in carbon with minute additions of titanium or vanadium, for example. Offering tensile strengths that can be triple the value of the carbon steels they are designed to replace (*e.g.*, 700 megapascals versus 200 megapascals), they have led to significant weight savings through thickness reductions—albeit at a slight loss of structural stiffness, because their elastic moduli are the same as other steels. They are considered to be quite competitive with aluminum substitutes for two reasons: they are relatively inexpensive (steel sells for one-half the price of aluminum on a per-unit-weight basis); and very little change in fabrication and processing procedures is needed in switching from carbon steel to HSLA steel, whereas major changes are usually required in switching to aluminum.

Bake-hardenable steels were developed specifically for the purpose of eliminating an expensive fabrication step—*i.e.*, the heat-treating furnace, where steels are imparted with their final strength. To do this, materials scientists have designed steels that can be strengthened in the same ovens used to bake body paint onto the part. These furnaces must operate at relatively low temperatures (170 °C, or 340 °F), so that special steels had to be developed that would achieve suitable strengths at heat-treatment temperatures very much below those normally employed (up to 600 °C, or 1,100 °F). Knowing that high-alloy steels would never be hardenable at such low temperatures, materials scientists focused their attention on carbon steels, but even here adequate strengths could not be obtained initially. Then in the 1980s scientists at the Japanese Sumitomo Metal Industries developed a steel containing nitrogen (a gas that constitutes three-quarters of the Earth's atmosphere) in addition to carbon and several other additives. Very high strengths (over 900 megapascals) and excellent toughness can be achieved on formed parts with this inexpensive addition after baking for 20 minutes at temperatures typical for a paint-baking operation.

Plastics and Composites

The motive for replacing the metal components of cars, trucks, and trains with plastics is the expectation of large weight savings due to the large differences in density involved: plastics are one-sixth the weight of steel and one-half that of aluminum per unit volume. However, as in evaluating the suitability of replacing steel with aluminum, the materials scientist must compare other properties of the materials in order to determine whether the tradeoffs are reasonable. For two reasons, the likely conclusion would be that plastics simply are not suitable for this type of application: the strength of most plastics, such as epoxies and polyesters, is roughly one-fifth that of steel or aluminum; and their elastic modulus is one-sixtieth that of steel and one-twentieth that of aluminum. On this basis, plastics do not appear to be suitable for structural components. What, then, accounts for the successful use that has been made of them? The answer lies in efforts made over the years by materials scientists, polymer chemists, mechanical engineers, and production managers to combine relatively weak and low-stiffness resins with high-strength, high-modulus reinforcements, thereby making new materials called composites with much more suitable properties than plastics alone.

The reinforcements used in composites are generally chosen for their high strength and modulus, as might be expected, but economic considerations often force compromises. For example, carbon fibres have extremely high modulus values (up to five times that of steel) and therefore make excellent reinforcements. However, their cost precludes their extensive use in automobiles, trucks, and trains, although they are used regularly in the aerospace industry. More suitable for non-aerospace applications are glass fibres (whose modulus can approach 1.5 times that of aluminum) or, in somewhat special cases, a mixture of glass and carbon fibres.

The physical form and shape of the reinforcements vary greatly, depending on many factors. The most effective reinforcements are long fibres, which are employed either in the form of a woven cloth or as separate layers of unidirectional fibres stacked upon one another until the proper laminate thickness is achieved. The resin may be applied to the fibres or cloth before laying up, thus forming what are termed prepregs, or it may be added later by "wetting out" the fibres. In either case, the assembly is then cured, usually under pressure, to form the composite. This type of composite takes full advantage of the properties of the fibres and is therefore capable of yielding strong, stiff panels. Unfortunately, the labour involved in the lay-up operations and other factors make it very expensive, so that long-fibre reinforcement is used only sparingly in the automobile industry.

One attempt to avoid expensive hand lay-up operations involves chopped fibres that are employed in mat form, somewhat like felt, or as loose fibres that may be either blown into a mold or injected into a mold along with the resin. Another method does not use fibres at all; instead the reinforcement is in the form of small, high-modulus particles. These are the least expensive of all to process, since the particles are simply mixed into the resin, and the mixture is used in various types of molds. On the other hand, particles are the least efficient reinforcement material; as a consequence, property improvements are not outstanding.

In choosing the other major constituent in composites, the polymer matrix, one faces a somewhat daunting variety, including epoxies, polyimides, polyurethanes, and polyesters. Each has its advantages and disadvantages that must be evaluated in order to determine suitability for a particular application. Among the factors to be considered are cost, processing temperature (curing temperature if using a thermoset polymer and melting temperature if using a thermoplastic), flow properties in the molding operation, sag resistance during paint bake out, moisture resistance, and shelf life. The number of combinations of resins, reinforcements, production methods, and fibre-to-resin ratios is so challenging that materials scientists must join forces with polymer chemists and engineers from the design, production, and quality-control departments of the company in order to choose the right combination for the application.

Judging by the inroads that have been made in replacing metals with composites, it appears that technologists have been making the right choices. The introduction of fibreglass-reinforced plastic skins on General Motors' 1953 Corvette sports car marked the first appearance of composites in a production model, and composites have continued to appear in automotive components ever since. In 1984, General Motors' Fiero was placed on the market with the entire body made from composites, and the Camaro/Firebird models followed with doors, roof panels, fenders, and other parts made of composites. Composites were also chosen for exterior panels in the Saturn, which appeared in 1990. In addition, they have had less visible applications—for example, the glass-reinforced nylon air-intake manifold on some BMW models.

Ceramics

Ceramics play an important role in engine efficiency and pollution abatement in automobiles and trucks. For example, one type of ceramic, cordierite (a magnesium aluminosilicate), is used as a substrate and support for catalysts in catalytic converters. It was chosen for this purpose because, along with many ceramics, it is lightweight, can operate at very high temperatures without melting, and conducts heat poorly (helping to retain exhaust heat for improved catalytic efficiency). In a novel application of ceramics, a cylinder wall was made of transparent sapphire (aluminum oxide) by General Motors' researchers in order to examine visually the internal workings of a gasoline engine combustion chamber. The intention was to arrive at improved understanding of combustion control, leading to greater efficiency of internal-combustion engines.

Another application of ceramics to automotive needs is a ceramic sensor that is used to measure the oxygen content of exhaust gases. The ceramic, usually zirconium oxide to which a small amount of yttrium has been added, has the property of producing a voltage whose magnitude depends on the partial pressure of oxygen surrounding the material. The electrical signal obtained from such a sensor is then used to control the fuel-to-air ratio in the engine in order to obtain the most efficient operation.

Because of their brittleness, ceramics have not been used as load-bearing components in ground-transportation vehicles to any great extent. The problem remains a challenge to be solved by materials scientists of the future.

Materials for Aerospace

The primary goal in the selection of materials for aerospace structures is the enhancement of fuel efficiency to increase the distance traveled and the payload delivered. This goal can be attained by developments on two fronts: increased engine efficiency through higher operating temperatures and reduced structural weight. In order to meet these needs, materials scientists look to materials in two broad areas—metal alloys and advanced composite materials. A key factor contributing to the advancement of these new materials is the growing ability to tailor materials to achieve specific properties.

Metals

Many of the advanced metals currently in use in aircraft were designed specifically for applications in gas-turbine engines, the components of which are exposed to high temperatures, corrosive gases, vibration, and high mechanical loads. During the period of early jet engines, design requirements were met by the development of new alloys alone. But the more severe requirements of advanced propulsion systems have driven the development of novel alloys that can withstand temperatures greater than 1,000 °C (1,800 °F), and the structural performance of such alloys has been improved by developments in the processes of melting and solidification.

Melting and Solidifying

Alloys are substances composed of two or more metals or of a metal and a nonmetal that are intimately united, usually by dissolving in each other when they are melted. The principal objectives of melting are to remove impurities and to mix the alloying ingredients homogeneously in the base metal. Major advances have been made with the development of new processes based on melting under vacuum (hot isostatic pressing), rapid solidification, and directional solidification.

In hot isostatic pressing, prealloyed powders are packed into a thin-walled, collapsible container, which is placed in a high-temperature vacuum to remove adsorbed gas molecules. It is then sealed and put in a press, where it is exposed to very high temperatures and pressures. The mold collapses and welds the powder together in the desired shape.

Molten metals cooled at rates as high as a million degrees per second tend to solidify into a relatively homogeneous microstructure, since there is insufficient time for crystalline grains to nucleate and grow. Such homogeneous materials tend to be stronger than the typical "grainy" metals. Rapid cooling rates can be achieved by "splat" cooling, in which molten droplets are projected onto a cold surface. Rapid heating and solidification can also be achieved by passing high-power laser beams over the material's surface.

Unlike composite materials, grainy metals exhibit properties that are essentially the same in all directions, so they cannot be tailored to match anticipated load paths (*i.e.,* stresses applied in specific directions). However, a technique called directional solidification provides a certain degree of tailorability. In this process the temperature of the mold is precisely controlled to promote the formation of aligned stiff crystals as the molten metal cools. These serve to reinforce the component in the direction of alignment in the same fashion as fibres reinforce composite materials.

Alloying

These advances in processing have been accompanied by the development of new "superalloys." Superalloys are high-strength, often complex alloys that are resistant to high temperatures and severe mechanical stress and that exhibit high surface stability. They are commonly classified into three major categories: nickel-based, cobalt-based, and iron-based. Nickel-based superalloys predominate in the turbine section of jet engines. Although they have little inherent resistance to oxidation at high temperatures, they gain desirable properties through the addition of cobalt, chromium, tungsten, molybdenum, titanium, aluminum, and niobium.

Aluminum-lithium alloys are stiffer and less dense than conventional aluminum alloys. They are also "superplastic," owing to the fine grain size that can now be achieved in processing. Alloys in this group are appropriate for use in engine components exposed to intermediate to high temperatures; they can also be used in wing and body skins.

Titanium alloys, as modified to withstand high temperatures, are seeing increased use in turbine engines. They are also employed in airframes, primarily for military aircraft but to some extent for commercial planes as well.

Composites

While developments in metals have had an impact on engine design, there is a growing trend toward the application of composite materials to aerospace structures. One of the reasons for this is that alloys do not offer substantial weight savings, which is a primary advantage of composites. Indeed, advanced composites have been used most widely where saving mass results in either significantly improved performance or significantly lower life-cycle costs. The most extensive application, therefore, has been in satellite systems, military aircraft, radomes, helicopters, commercial transport aircraft, and general aviation.

Broadly defined, composites are materials with two or more distinct components that combine to yield characteristics superior to those of the individual constituents. Although this definition can apply to such ordinary building materials as plywood, concrete, and bricks, within the aerospace industry the term composite generally refers to the fibre-reinforced metal, polymer, and ceramic products that have come into use since World War II. These materials consist of fibres (such as glass, graphite, silicon carbide, or aramid) that are embedded in a matrix of, for example, aluminum, epoxy, or silicon nitride.

In the late 1950s a revolution in materials development occurred in response to the space program's need for lightweight, thermally stable materials. Boron-tungsten filaments, carbon-graphite fibres, and organic aramid fibres proved to be strong, stiff, and light, but one problem with using them as fibres was that they were of limited value in any construction other than rope, which can bear loads in only one direction. Materials scientists needed to develop a way to make them useful under all loading conditions, and this led to the development of composites. While the structural value of a bundle of fibres is low, the strength of individual fibres can be harnessed if they are embedded in a matrix that acts as an adhesive, binding the fibres and lending solidity to the material. The matrix also protects the fibres from environmental stress and physical damage, which can initiate cracks. In addition, while the strength and stiffness of the composite remain largely a function of the reinforcing material—that is, the fibres—the matrix can contribute other properties, such as thermal and electrical conductivity and, most important, thermal stability. Finally, fibre-matrix combination reduces the potential for complete fracture. In a monolithic (or single) material, a crack, once started, generally continues to propagate until the material fails; in a composite, if one fibre in an assemblage fails, the crack may not extend to the other fibres, so the damage is limited.

To some extent, the composite-materials engineer is trying to mimic structures made spontaneously by plants and animals. A tree, for example, is made of a fibre-reinforced material whose strength is derived from cellulose fibres that grow in directions that match the weight of the branches. Similarly, many organisms naturally fabricate "bioceramics," such as those found in shells, teeth, and bones. While the designers of composites for the aerospace industry would like to copy some of the features of bioceramics production—room-temperature processing and net-shape products, for example—they do not want to be constrained by slow processing methods and limited fibre and matrix material choices. In addition, unlike a mollusk, which has to produce only one shell, the composites manufacturer has to use rapid, repeatable processing methods that can fabricate hundreds or even thousands of parts.

Modern composites are generally classified into three categories according to the matrix material: polymer, metal, or ceramic. Since polymeric materials tend to degrade at elevated temperatures, polymer-matrix composites (PMCs) are restricted to secondary structures in which operating temperatures are lower than 300 °C (570 °F). For higher temperatures, metal-matrix and ceramic-matrix composites are required.

Polymer-Matrix Composites

PMCs are of two broad types, thermosets and thermoplastics. Thermosets are solidified by irreversible chemical reactions, in which the molecules in the polymer "cross-link," or form connected chains. The most common thermosetting matrix materials for high-performance composites used in the aerospace industry are the epoxies. Thermoplastics, on the other hand, are melted and then solidified, a process that can be repeated numerous times for reprocessing. Although the manufacturing technologies for thermoplastics are generally not as well developed as those for thermosets, thermoplastics offer several advantages. First, they do not have the shelf-life problem associated with thermosets, which require freezer storage to halt the irreversible curing process that begins at room temperature. Second, they are more desirable from an environmental point of view, as they can be recycled. They also exhibit higher fracture toughness and better resistance to solvent attack. Unfortunately, thermoplastics are more expensive, and they generally do not resist heat as well as thermosets; however, strides are being made in developing thermoplastics with higher melting temperatures. Overall, thermoplastics offer a greater choice of processing approaches, so that the process can be determined by the scale and rate of production required and by the size of the component.

A variety of reinforcements can be used with both thermoset and thermoplastic PMCs, including particles, whiskers (very fine single crystals), discontinuous (short) fibres, continuous fibres, and textile preforms (made by braiding, weaving, or knitting fibres together in specified designs). Continuous fibres are more efficient at resisting loads than are short ones, but it is more difficult to fabricate complex shapes from materials containing continuous fibres than from short-fibre or particle-reinforced materials. To aid in processing, most high-performance composites are strengthened with filaments that are bundled into yarns. Each yarn, or tow, contains thousands of filaments, each of which has a diameter of approximately 10 micrometres (0.01 millimetre, or 0.0004 inch).

Depending on the application and on the type of load to be applied to the composite part, the reinforcement can be random, unidirectional (aligned in a single direction), or multidirectional (oriented in two or three dimensions). If the load is uniaxial, the fibres are all aligned in the load direction to gain maximum benefit of their stiffness and strength. However, for multidirectional loading (for example, in aircraft skins), the fibres must be oriented in a variety of directions. This is often accomplished by stacking layers (or lamina) of continuous-fibre systems.

The most common form of material used for the fabrication of composite structures is the preimpregnated tape, or "prepreg." There are two categories of prepreg: tapes, generally 75 millimetres (3 inches) or less in width, intended for fabrication in automated, computer-controlled tape-laying machines; and "broad goods," usually several metres in dimension, intended for hand lay-up and large sheet applications. To make prepregs, fibres are subjected to a surface treatment so that the resin will adhere to them. They are then placed in a resin bath and rolled into tapes or sheets.

To fabricate the composite, the manufacturer "lays up" the prepreg according to the reinforcement needs of the application. This has traditionally been done by hand, with successive layers of a broad-goods laminate stacked over a tool in the shape of the desired part in such a way as to accommodate the anticipated loads. However, efforts are now being directed toward automated fibre-placement methods in order to reduce costs and ensure quality and repeatability. Automated fibre-placement processes fall into two categories, tape laying and filament winding. The tape-laying process involves the use of devices that control the placement of narrow prepreg tapes over tooling with the contours of the desired part and along paths prescribed by the design requirements of the structure. The width of the tape determines the "sharpness" of the turns required to place the fibres in the prescribed direction–i.e., wide tapes are used for gradual turns, while narrow tapes are required for the sharp turns associated with more complex shapes.

Filament winding uses the narrowest prepreg unit available–the yarn, or tow, of impregnated filaments. In this process, the tows are wound in prescribed directions over a rotating mandrel in the shape of the part. Successive layers are added until the required thickness is reached. Although filament winding was initially limited to geodesic paths (*i.e.*, winding the fibres along the most direct route between two points), the process is now capable of fabricating complex shapes through the use of robots.

For thermosetting polymers, the structure generated by either tape laying or filament winding must undergo a second manipulation in order to solidify the polymer through a curing reaction.

This is usually accomplished by heating the completed structure in an autoclave, or oven. Thermoplastic systems offer the advantage of on-line consolidation, so that the high energy and capital costs associated with the curing step can be eliminated. For these systems, prepreg can be locally melted, consolidated, and cooled at the point of contact so that a finished structure is produced. A variety of energy sources are used to concentrate heat at the point of contact, including hot-gas torches, infrared light, and laser beams.

Pultrusion, the only truly continuous process for manufacturing parts from PMCs, is economical but limited to the production of beamlike shapes. On a pultrusion line, fibres and the resin are pushed through a heated die, or shaping tool, at one end, then cooled and pulled out at the other end. This process can be applied to both thermoplastic and thermoset polymers.

Resin transfer molding, or RTM, is a composites processing method that offers a high potential for tailorability but is currently limited to low-viscosity (easily flowing) thermosetting polymers. In RTM, a textile preform—made by braiding, weaving, or knitting fibres together in a specified design—is placed into a mold, which is then closed and injected with a resin. After consolidation, the mold is opened and the part removed. Preforms can be made in a wide variety of architectures, and several can be joined together during the RTM process to form a multi-element preform offering reinforcement in specific areas and load directions.

The similarity of meltable thermoplastic polymers to metals has prompted the extension of techniques used in metalworking. Sheet forming, used since the 19th century by metallurgists, is now applied to the processing of thermoplastic composites. In a typical thermoforming process, the sheet stock, or preform, is heated in an oven. At the forming temperature, the sheet is transferred into a forming system, where it is forced to conform to a tool, with a shape that matches the finished part. After forming, the sheet is cooled under pressure and then removed. Stretch forming, a variation on thermoplastic sheet forming, is specifically designed to take advantage of the extensibility, or ability to be stretched, of thermoplastics reinforced with long, discontinuous fibres. In this process, a straight preconsolidated beam is heated and then stretched over a shaped tool to introduce curvature. The specific advantage of stretch forming is that it provides an automated way to achieve a very high degree of fibre-orientation control in a wide range of part sizes.

Metal-Matrix and Ceramic-Matrix Composites

The requirement that finished parts be able to operate at temperatures high enough to melt or degrade a polymer matrix creates the need for other types of matrix materials, often metals. Metal matrices offer not only high-temperature resistance but also strength and ductility, or "bendability," which increases toughness. The main problems with metal-matrix composites (MMCs) are that even the lightest metals are heavier than polymers, and they are very complex to process. MMCs can be used in such areas as the skin of a hypersonic aircraft, but on wing edges and in engines temperatures often exceed the melting point of metals. For the latter applications, ceramic-matrix composites (CMCs) are seeing increasing use, although the technology for CMCs is less mature than that for PMCs. Ceramics consist of alumina, silica, zirconia, and other elements refined from fine earth and sand or of synthetic materials, such as silicon nitride or silicon carbide. The desirable properties of ceramics include superior heat resistance and low abrasive and corrosive properties. Their primary drawback is brittleness, which can be reduced by reinforcing with fibres or whiskers. The reinforcement material can be a metal or another ceramic.

Unlike polymers and metals, which can be processed by techniques, that involve melting (or softening) followed by solidification, high-temperature ceramics cannot be melted. They are generally produced by some variation of sintering, a technique that renders a combination of materials into a coherent mass by heating to high temperatures without complete melting. If continuous fibres or textile weaves (as opposed to short fibres or whiskers) are involved, sintering is preceded by impregnating the assembly of fibres with slurry of ceramic particles dispersed in a liquid. A major benefit of using CMCs in aircraft engines is that they allow higher operating temperatures and thus greater combustion efficiency, leading to reduced fuel consumption. An additional benefit is derived from the low density of CMCs, which translates into substantial weight savings.

Other Advanced Composites

Carbon-carbon composites are closely related to CMCs but differ in the methods by which they are produced. Carbon-carbon composites consist of semicrystalline carbon fibres embedded in a matrix of amorphous carbon. The composite begins as a PMC, with semicrystalline carbon fibres impregnated with a polymeric phenolic resin. The resin-soaked system is heated in an inert atmosphere to pyrolyze, or char, the polymer to a carbon residue. The composite is re-impregnated with polymer, and the pyrolysis is repeated. Continued repetition of this impregnation/pyrolysis process yields a structure with minimal voids. Carbon-carbon composites retain their strength at 2,500 °C (4,500 °F) and are used in the nose cones of reentry vehicles. However, because they are vulnerable to oxidation at such high temperatures, they must be protected by a thin layer of ceramic.

While materials research for aerospace applications has focused largely on mechanical properties such as stiffness and strength, other attributes are important for use in space. Materials are needed with a near-zero coefficient of thermal expansion; in other words, they have to be thermally stable and should not expand and contract when exposed to extreme changes in temperature. A great deal of research is focused on developing such materials for high-speed civilian aircraft, where thermal cycling is a major issue. High-toughness materials and non-flammable resin composite systems are also under investigation to improve the safety of aircraft interiors.

Efforts are also being directed toward the development of "smart," or responsive, materials. Representing another attempt to mimic certain characteristics of living organisms, smart materials, with their built-in sensors and actuators, would react to their external environment by bringing on a desired response. This would be done by linking the mechanical, electrical, and magnetic properties of these materials. For example, piezoelectric materials generate an electrical current when they are bent; conversely, when an electrical current is passed through these materials, they stiffen. This property can be used to suppress vibration: the electrical current generated during vibration could be detected, amplified, and sent back, causing the material to stiffen and stop vibrating.

Materials for Computers and Communications

The basic function of computers and communications systems is to process and transmit information in the form of signals representing data, speech, sound, documents, and visual images. These signals are created, transmitted, and processed as moving electrons or photons, and so the basic materials groups involved are classified as electronic and photonic. In some cases, materials known as optoelectronic bridge these two classes, combining abilities to interact usefully with both electrons and photons.

Among the electronic materials are various crystalline semiconductors; metalized film conductors; dielectric films; solders; ceramics and polymers formed into substrates on which circuits are assembled or printed; and gold or copper wiring and cabling.

Photonic materials include a number of compound semiconductors designed for light emission or detection; elemental dopants that serve as photonic performance-control agents; metal- or diamond-film heat sinks; metalized films for contacts, physical barriers, and bonding; and silica glass, ceramics, and rare earths for optical fibres.

Electronic Materials

Between 1955 and 1990, improvements and innovations in semiconductor technology increased the performance and decreased the cost of electronic materials and devices by a factor of one million—an achievement unparalleled in the history of any technology. Along with this extraordinary explosion of technology has come an exponentially upward spiral of the capital investment necessary for manufacturing operations. In order to maintain cost-effectiveness and flexibility, radical changes in materials and manufacturing operations will be necessary.

Semiconductor Crystals

Silicon

Bulk semiconductor silicon for the manufacture of integrated circuits (sometimes referred to as electronic-grade silicon) is the purest material ever made commercially in large quantities. One of the most important factors in preparing this material is control of such impurities as boron, phosphorus, and carbon. For the ultimate levels of integrated-circuit design, stray contaminant atoms must constitute less than 0.1 part per trillion of the material.

For fabrication into integrated circuits, bulk semiconductor silicon must be in the form of a single-crystal material with high crystalline perfection and the desired charge-carrier concentration. The size of the silicon ingot, or boule, has been scaled up in recent years, in order to provide wafers of increasing diameter that are demanded by the economics of integrated-circuit manufacturing. Most commonly, a 60-kilogram (130-pound) charge is grown to an ingot with a diameter of 200 millimetres (8 inches), but the semiconductor industry will soon require ingots as large as 300 millimetres. The ingots are then converted into wafers by machining and chemical processes.

III–V Compounds

Although silicon is by far the most commonly used crystal material for integrated circuits, a significant volume of semiconductor devices and circuits employs III–V technology, so named because it is based on crystalline compounds formed by combining metallic elements from column III and non-metallic elements from column V of the periodic table of chemical elements. When the elements are gallium and arsenic, the semiconductor is called gallium arsenide, or GaAs. However, other elements such as indium, phosphorus, and aluminum are often used in the compound to achieve specific performance characteristics.

For electronic applications, the III–V semiconductors offer the basic advantage of higher electron mobility, which translates into higher operating speeds. In addition, devices made with III–V compounds provide lower voltage operation for specific functions, radiation hardness (especially important for satellites and space vehicles), and semi-insulating substrates (avoiding the presence of parasitic capacitance in switching devices).

III–V materials are more difficult to handle than silicon, and a III–V wafer or substrate usually is less than half the size of a silicon wafer. In addition, a gallium arsenide wafer entering the processing facility can be expected to cost 10 to 20 times as much as a silicon wafer, although that cost difference narrows somewhat after fabrication, packaging, and testing. Nevertheless, there is one major characteristic of III–V materials with which silicon cannot compete: a III–V compound can be tailored to generate or detect photons of a specific wavelength. For example, an indium gallium arsenide phosphide (InGaAsP) laser can generate radiation at 1.55 micrometres to carry digitally coded information streams. This means that a III–V component can fill both electronic and photonic functions in the same integrated circuit.

Photoresist Films

Patterning polished wafers with an integrated circuit requires the use of photoresist materials that form thin coatings on the wafer before each step of the photolithographic process. Modern photoresists are polymeric materials that are modified when exposed to radiation (either in the form of visible, ultraviolet, or X-ray photons or in the form of energetic electron beams). A photoresist typically contains a photoactive compound (PAC) and an alkaline-soluble resin. The PAC, mixed into the resin, renders it insoluble. This mixture is coated onto the semiconductor wafer and is then exposed to radiation through a "mask" that carries the desired pattern. Exposed PAC is converted into an acid that renders the resin soluble, so that the resist can be dissolved and the exposed substrate beneath it chemically etched or metallically coated to match the circuit design.

Besides practical properties such as shelf life, cost, and availability, the key properties of a photoresist include purity, etching resistance, resolution, contrast, and sensitivity. As the feature sizes of integrated circuits shrink in each successive generation of microchips, photoresist materials are challenged to handle shorter wavelengths of light. For example, the photolithography of current designs (with features that have shrunk to less than one micrometre) is based on ultraviolet radiation in the wavelength range of 365 to 436 nanometres, but, in order to define accurately the smaller features of future microchips (less than 0.25 micrometre), shorter wavelengths will be necessary. The problem here is that electromagnetic radiation in such frequency regions is weaker. One solution is to use the chemically amplified photoresist, or CAMP. The sensitivity of a photoresist is measured by its quantum efficiency, or the number of chemical events that occur when a photon is absorbed by the material. In CAMP material, the number of events is dramatically increased by subsequent chemical reactions (hence the amplification), which means that less light is needed to complete the process.

Electric Connections

The performance of today's electronic systems (and photonic systems as well) is limited significantly by interconnection technology, in which components and subsystems are linked by conductors and connectors. Currently, very fine gold or copper wiring, as thin as 30 micrometres, is used to carry electric current to and from the many pads along the sides or ends of a microchip to other components on a circuit board. The capacitance involved in such circuitry slows down the flow of electrons and, hence, of information. However, by integrating several chips into a single multichip module, in which the chips are connected on a shared substrate by various conducting materials (such as metalized film), the speed of information flow, can be increased, thus improving the assembly's performance. Ideally, all the chips in a single module would be fabricated simultaneously on the same wafer, but in practice this is not feasible: Silicon crystal manufacture is still subject to an average of one flaw per wafer, meaning that at least one of the many chips cut from each wafer is scrapped. If the whole wafer area were dedicated to a single multifunction assembly, that one flaw would scrap the entire module. Multichip modules are therefore made up of as many as five microchips bonded to a silicon or ceramic substrate on which resistors and capacitors have been constructed with thin films. Typical materials used in a multichip module include the substrate; gold paste conductors applied in an additive process resembling silk screen printing; vitreous glazes to insulate the gold paste conductors from subsequent film layers; a series of thin films made with tantalum nitride, titanium, palladium, and plated gold; and a final package of silicone rubber.

Packaging Materials

Several major types of packaging material are used by the electronics industry, including ceramic, refractory glass, premolded plastic, and postmolded plastic. Ceramic and glass packages cost more than plastic packages, so they make up less than 10 percent of the worldwide total. However, they provide the best protection for complex chips. Premolded plastic packages account for only a small but important fraction of the market, since they are required for packaging devices with many leads. Most plastic packages are postmolded, meaning that the package body is molded over the assembly after the microchip has been attached to the fan-out pattern.

Precursors

The starting materials for most semiconductor devices are volatile and ultrapure gaseous derivatives of various organic and inorganic precursors. Many of them are toxic, and many will ignite spontaneously in the atmosphere. These gases are transported in high-pressure cylinders from the plant where they were made to the site where they will be used. One possible method of replacing these precursors with materials that are environmentally safe is known as in situ synthesis. In this method, dangerous reagents would be generated on demand in only the desired quantities, instead of being shipped cross-country and stored until needed at the semiconductor processing plant.

Photonic Materials

Computers and communications systems have been dominated by electronic technology since their beginnings, but photonic technology is making serious inroads throughout the information movement and management systems with such devices as lasers, light-emitting diodes, photo-detecting diodes, optical switches, optical amplifiers, optical modulators, and optical fibres. Indeed, for long-distance terrestrial and transoceanic transmission of information, photonics has almost completely displaced electronics.

Crystalline Materials

The light detectors and generators listed above are actually optoelectronic, because they link photonic and electronic systems. They employ the III–V compound semiconductors many of them characterized by their band gaps—*i.e.,* the energy minimum of the electron conduction band and the energy maximum of hole valence bands occur at the same location in the momentum space, allowing electrons and holes to recombine and radiate photons efficiently. (By contrast, the conduction band minimum and the valence band maximum in silicon have dissimilar momenta, and therefore the electrons and holes cannot recombine efficiently.) Among the important compounds are gallium arsenide, aluminum gallium arsenide, indium gallium arsenide phosphide, indium phosphide, and aluminum indium arsenide.

Fabricating a single crystal from these combinations of elements is far more difficult than creating a single crystal of electronic-grade silicon. Special furnaces are required, and the process can take several days. Notwithstanding the precision involved, the sausage-shaped boule is less than half the diameter of a silicon ingot and is subject to a much higher rate of defects. Researchers are continuously seeking ways to reduce the thermal stresses that are primarily responsible for dislocations in the III–V crystal lattice that cause these defects. The purity and structural perfection of the final single-crystal substrates affect the qualities of the crystalline layers that are grown on them and the regions that are diffused or implanted in them during the manufacture of photonic devices.

Epitaxial Layers

For the efficient emission or detection of photons, it is often necessary to constrain these processes to very thin semiconductor layers. These thin layers, grown atop bulk semiconductor wafers, are called epitaxial layers because their crystallinity matches that of the substrate even though the composition of the materials may differ—e.g., gallium aluminum arsenide (GaAlAs) grown atop a gallium arsenide substrate. The resulting layers form what is called a heterostructure. Most continuously operating semiconductor lasers consist of heterostructures, a simple example consisting of 1000-angstrom thick gallium arsenide layers sandwiched between somewhat thicker (about 10000 angstroms) layers of gallium aluminum arsenide—all grown epitaxially on a gallium arsenide substrate. The sandwiching and repeating of very thin layers of a semiconductor between layers of a different composition allow one to modify the band gap of the sandwiched layer. This technique, called band-gap engineering, permits the creation of semiconductor materials with properties that cannot be found in nature. Band-gap engineering, used extensively with III–V compound semiconductors, can also be applied to elemental semiconductors such as silicon and germanium.

The most precise method of growing epitaxial layers on a semiconducting substrate is molecular-beam epitaxy (MBE). In this technique, a stream or beam of atoms or molecules is effused from a common source and travels across a vacuum to strike a heated crystal surface, forming a layer that has the same crystal structure as the substrate. Variations of MBE include elemental-source MBE, hydride-source MBE, gas-source MBE, and metal-organic MBE. Other approaches to epitaxial growth are liquid-phase epitaxy (LPE) or chemical vapour deposition (CVD). The latter method includes hydride CVD, trichloride CVD, and metal-organic CVD.

Normally, epitaxial layers are grown on flat surfaces, but scientists are searching for an economical and reliable method of growing epitaxial material on nonplanar structures—for example, around the "mesas" or "ridges" or in the "tubs" or "channels" that are etched into the surface of semiconducting devices. Nonplanar epitaxy is considered necessary for producing monolithic integrated optical devices or all-photonic switches and logic elements, but mastery of this method requires better understanding of the surface chemistry and surface dynamics of epitaxial growth.

Optical Switching

Research in this area is driven by the need to switch data streams of higher and higher speed efficiently as customers for computer and communications services demand transmission and switching rates far higher than can be provided by a purely electronic system. Thanks to developments in semiconductor lasers and detectors and in optical fibres, transmission at the desired high speeds has become possible. However, the switching of optical data streams still requires converting the data from the optical to the electronic domain, subjecting them to electronic switching and to manipulation inside the switching apparatus, and then reconverting the switched and reconfigured data into the optical domain for transmission over optical fibres. Electronic switching therefore is seen as the principal barrier to achieving higher switching speeds. One approach to solving this problem would be to introduce optics inside digital switching machines. Known as free-space photonics, this approach would involve such devices as semiconductor lasers or light-emitting diodes (LEDs), optical modulators, and photo detectors—all of which would be integrated into systems combined with electronic components.

One commercially available device for photonic switching is the quantum-well self-electro-optic-effect device, or seed. The key concept for this device is the use of quantum wells. These structures consist of many thin layers of two different semiconductor materials. Individual layers are typically 10 nanometres (about 40 atoms) thick and 100 layers are used in a device about 1 micrometre thick. When a voltage is applied across the layers, the transmission of photons through the quantum wells changes significantly, in effect creating an optical modulator—an essential component of any photonic circuit. Variations on the SEED concept are the symmetric SEED (S-SEED) and the field-effect transistor SEED. Neighbouring S-SEEDs could be connected by pairs of back-to-back quantum-well photodiodes, and commercially sized interconnection networks could be built by using free-space photonic interconnections between two-dimensional arrays of switching nodes. However, even this type of free-space optical interconnection technology would only enhance and extend electronic technology, not replace it.

The move of optoelectronic and photonic integrated circuits out of the research laboratory and into the marketplace has been made possible by the availability of high-quality epitaxial growth techniques for building up lattice-matched crystalline layers of indium gallium arsenide phosphide and indium phosphide (InGaAsP/InP). This III–V compound system is central to the light emitters and detectors used in the 1.3-micrometre and 1.5-micrometre wavelength ranges at which optical fibre has very low transmission loss.

Optical Transmission

As the rates of transmission are increased from millions of bits (megabits) per second to billions of bits (gigabits) per second, commercially available lasers encounter a physical limitation called "chirping," in which the optical frequency of the laser begins to waver during a pulse. Future systems, which may require from 2.4 to 30 gigabits per second, are probably going to be based on the use of a continuously operating distributed-feedback laser, whose output will be modulated in intensity by passing it through a modulator. This device consists of a crystal substrate of lithium niobate onto which a titanium channel is diffused to function as a light guide. The signal is encoded onto the light beam via a microwave radio-frequency feed through neighbouring channels in the coupler. Such a device is used only at the transmitter end of the optical path.

Both communications and computer systems rely on silica glass fibres to transmit light signals from lasers and LEDs. For long-distance transmission, optical-fibre cables are usually equipped with electro-optical repeater assemblies approximately every 100 kilometres. A new approach, called optical amplifiers, has been developed for deployment in transoceanic fibre-optic cables. Unlike traditional repeaters, optical amplifiers work by adding photons to a light signal without changing it to an electrical signal and without changing its bit-rate. Since they can be used at any desired transmission bit-rate, a transoceanic cable equipped with these devices can be upgraded to higher bit-rates simply by changing the lasers and photodiodes at each end. No retrofitting of higher bit-rate amplifiers is necessary.

The optical amplifier is a module containing a semiconductor pump laser and a short length of optical fibre whose core has been doped with less than 0.1 percent erbium, an optically active rare-earth element. The pump laser is powered by an electrical conductor that runs the length of the cable. The amplifier functions by converting the optical energy generated by the pump source into signal photon energy. When a signal-carrying stream of laser pulses passes through the optical amplifier, it is combined with the pump light through a wavelength division multiplexer located in the module. The combined signal is fed through the erbium-doped fibre length, where the excited erbium ions contribute photons coherently to the signal.

Materials for Medicine

The treatment of many human disease conditions requires surgical intervention in order to assist, augment, sustain, or replace a diseased organ, and such procedures involve the use of materials foreign to the body. These materials, known as biomaterials, include synthetic polymers and, to a lesser extent, biological polymers, metals, and ceramics. Specific applications of biomaterials range from high-volume products such as blood bags, syringes, and needles to more challenging implantable devices designed to augment or replace a diseased human organ. The latter devices are used in cardiovascular, orthopedic, and dental applications as well as in a wide range of invasive treatment and diagnostic systems. Many of these devices have made possible notable clinical successes. For example, in cardiovascular applications, thousands of lives have been saved by heart valves, heart pacemakers, and large-diameter vascular grafts and orthopedic hip-joint replacements have shown great long-term success in the treatment of patients suffering from debilitating joint diseases. With such a tremendous increase in medical applications, demand for a wide range of biomaterials grows by 5 to 15 percent each year. In the United States the annual market for surgical implants exceeds $10 billion, approximately 10 percent of world demand.

Nevertheless, applications of biomaterials are limited by biocompatibility, the problem of adverse interactions arising at the junction between the biomaterial and the host tissue. Optimizing the interactions that occur at the surface of implanted biomaterials represents the most significant key to further advances, and an excellent basis for these advances can be found in the growing understanding of complex biological materials and in the development of novel biomaterials custom-designed at the molecular level for specific medical applications.

General Requirements of Biomaterials

Research on developing new biomaterials is an interdisciplinary effort, often involving collaboration among materials scientists and engineers, biomedical engineers, pathologists, and clinicians to solve clinical problems. The design or selection of a specific biomaterial depends on the relative importance of the various properties that are required for the intended medical application. Physical properties that are generally considered include hardness, tensile strength, modulus, and elongation; fatigue strength, which is determined by a material's response to cyclic loads or strains; impact properties; resistance to abrasion and wear; long-term dimensional stability, which is described by a material's viscoelastic properties; swelling in aqueous media; and permeability to gases, water, and small biomolecules. In addition, biomaterials are exposed to human tissues and fluids, so that predicting the results of possible interactions between host and material is an important and unique consideration in using synthetic materials in medicine. Two particularly important issues in biocompatibility are thrombosis, which involves blood coagulation and the adhesion of blood platelets to biomaterial surfaces, and the fibrous-tissue encapsulation of biomaterials that are implanted in soft tissues.

Poor selection of materials can lead to clinical problems. One example of this situation was the choice of silicone rubber as a poppet in an early heart valve design. The silicone absorbed lipid from plasma and swelled sufficiently to become trapped between the metal struts of the valve. Another unfortunate choice as a biomaterial was Teflon (trademark), which is noted for its low coefficient of friction and its chemical inertness but which has relatively poor abrasion resistance. Thus, as an occluder in a heart valve or as an acetabular cup in hip-joint prosthesis, Teflon may eventually wear to such an extent that the device would fail. In addition, degradable polyesterurethane foam was abandoned as a fixation patch for breast prostheses, because it offered a distinct possibility for the release of carcinogenic by-products as it degraded.

Besides their constituent polymer molecules, synthetic biomaterials may contain several additives, such as unreacted monomers and catalysts, inorganic fillers or organic plasticizers, antioxidants and stabilizers, and processing lubricants or mold-release agents on the material's surface. In addition, several degradation products may result from the processing, sterilization, storage, and ultimately implantation of a device. Many additives are beneficial—for example, the silica filler that is indispensable in silicone rubber for good mechanical performance or the antioxidants and stabilizers that prevent premature oxidative degradation of polyetherurethanes. Other additives, such as pigments, can be eliminated from biomedical products. Indeed, a "medical-grade" biomaterial is one that has had nonessential additives and potential contaminants excluded or eliminated from the polymer. In order to achieve this grade, the polymer may need to be solvent-extracted before use, thereby eliminating low-molecular-weight materials. Generally, additives in polymers are regarded with extreme suspicion, because it is often the additives rather than the constituent polymer molecules that are the source of adverse biocompatibility.

Polymer Biomaterials

The majority of biomaterials used in humans are synthetic polymers such as the polyurethanes or Dacron, rather than polymers of biological origin such as proteins or polysaccharides. The properties of common synthetic biomaterials vary widely, from the soft and delicate water-absorbing hydrogels made into contact lenses to the resilient elastomers found in short- and long-term cardiovascular devices or the high-strength acrylics used in orthopedics and dentistry. The properties of any material are governed by its chemical composition and by the intra- and intermolecular forces that dictate its molecular organization. Macromolecular structure in turn affects macroscopic properties and, ultimately, the interfacial behaviour of the material in contact with blood or host tissues.

Since the properties of each material are dependent on the chemical structure and macromolecular organization of its polymer chains, an understanding of some common structural features of various polymers provides considerable insight into their properties. Compared with complex biological molecules, synthetic polymers are relatively simple; often they comprise only one type of repeating subunit, analogous to a polypeptide consisting of just one repeating amino acid. On the basis of common structures and properties, synthetic polymers are classified into one of three categories: elastomers, which include natural and synthetic rubbers; thermoplastics; and thermosets. The properties that provide the basis for this classification include molecular weight, cross-link density, percent crystallinity, thermal transition temperature, and bulk mechanical properties.

Elastomers

Elastomers, which include rubber materials, have found wide use as biomaterials in cardiovascular and soft-tissue applications owing to their high elasticity, impact resistance, and gas permeability. Applications of elastomers include flexible tubing for pacemaker leads, vascular grafts, and catheters; biocompatible coatings and pumping diaphragms for artificial hearts and left-ventricular assist devices; grafts for reconstructive surgery and maxillofacial operations; wound dressings; breast prostheses; and membranes for implantable biosensors.

Elastomers are typically amorphous with low cross-link density (although linear polyurethane block copolymers are an important exception). This gives them low to moderate modulus and tensile properties as well as high elasticity. For example, elastomeric devices can be extended by 100 to 1,000 percent of their initial dimensions without causing any permanent deformation to the material. Silicone rubbers such as Silastic (trademark), produced by the American manufacturer Dow Corning, Inc., are cross-linked, so that they cannot be melted or dissolved—although swelling may occur in the presence of a good solvent. Such properties contrast with those of the linear polyurethane elastomers, which consist of soft polyether amorphous segments and hard urethane-containing glassy or crystalline segments. The two segments are incompatible at room temperature and undergo microphase separation, forming hard domains dispersed in an amorphous matrix. A key feature of this macromolecular organization is that the hard domains serve as physical cross-links and reinforcing filler. This results in elastomeric materials that possess relatively high modulus and extraordinary long-term stability under sustained cyclic loading. In addition, they can be processed by methods common to thermoplastics.

Thermoplastics

Many common thermoplastics, such as polyethylene and polyester, are used as biomaterials. Thermoplastics usually exhibit moderate to high tensile strength (5 to 1,000 megapascals) with moderate elongation (2 to 100 percent), and they undergo plastic deformation at high strains.

Thermoplastics consist of linear or branched polymer chains; consequently, most can undergo reversible melt-solid transformation on heating, which allows for relatively easy processing or reprocessing. Depending on the structure and molecular organization of the polymer chains, thermoplastics may be amorphous (*e.g.*, polystyrene), semicrystalline (*e.g.*, low-density polyethylene), or highly crystalline (*e.g.*, high-density polyethylene), or they may be processed into highly crystalline textile fibres (*e.g.*, polyester Dacron).

Some thermoplastic biomaterials, such as polylactic acid and polyglycolic acid, are polymers based on a repeating amino acid subunit. These polypeptides are biodegradable, and, along with biodegradable polyesters and polyorthoesters, they have applications in absorbable sutures and drug-release systems. The rate of biodegradation in the body can be adjusted by using copolymers. These are polymers that link two different monomer subunits into a single polymer chain. The resultant biomaterial exhibits properties, including biodegradation, that are intermediate between the two homopolymers.

Thermosets

Thermosetting polymers find only limited application in medicine, but their characteristic properties, which combine high strength and chemical resistance, are useful for some orthopedic and dental devices. Thermosetting polymers such as epoxies and acrylics are chemically inert, and they also have high modulus and tensile properties with negligible elongation (1 to 2 percent).

The polymer chains in these materials are highly cross-linked and therefore have severely restricted macromolecular mobility; this limits extension of the polymer chains under an applied load. As a result, thermosets are strong but brittle materials.

Cross-linking inhibits close packing of polymer chains, preventing formation of crystalline regions. Another consequence of extensive cross-linking is that thermosets do not undergo solid-melt transformation on heating, so that they cannot be melted or reprocessed.

Materials Science and Engineering

Material science and engineering is a multidisciplinary field that combines engineering, physics and chemistry principles to solve real-world problems associated with nanotechnology, biotechnology, information technology, energy, manufacturing and other major engineering disciplines.

Materials science is an interdisciplinary field involving the study of different types of materials and the applications of knowledge about these materials to various areas of science and engineering. It combines elements of applied physics and chemistry, as well as chemical, mechanical, civil and electrical engineering. Materials science and materials engineering are often combined into a larger field of study.

Furthermore, materials science has grown to include testing these more exotic forms of condensed matter and developing new physics theories to explain their behavior. Consequently, materials science has been propelled to the forefront at many academic institutions and research facilities.

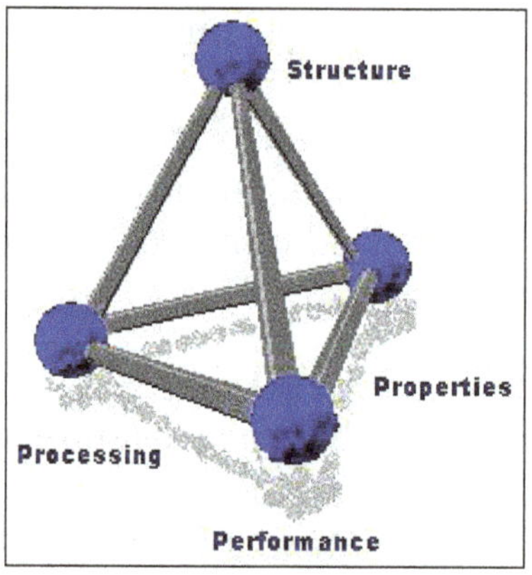

The Materials Science Tetrahedron shows the four main areas in which materials are studied. It often includes Characterization at the center.

Materials research at the basic level can lead to unprecedented influence on society. For example, semiconductor materials, which are ubiquitous in cars, telephones, computers, clocks, kitchen appliances, children's toys, satellites, telescopes, and more, were a product of materials science research—into the electronic properties of the element germanium. Further research led to the replacement of germanium with the less costly silicon and to diverse approaches to modifying silicon's properties by implanting other elements, such as phosphorous or boron, into the silicon matrix. Since their discovery in 1947, semiconductors have been steadily improved through materials science research driven by ever-increasing performance demands from the computer industry.

Efforts to apply ethical considerations to Materials Science quickly reach what is a common barrier between ethics and the combined fields of science and technology. An individual scientist, for example, who would want to conduct research toward such a noble goal as developing a light-weight and durable structural plastic that is readily recyclable must first either find and join a research group that is already funded to support such research or find an independent funding source for such research.

Fundamentals of Materials Science

In materials science, the researcher conducts a systematic investigation of each material, in terms of its structure, properties, processing, and performance. The research often leads to new applications of known materials and the creation of new materials with desired properties.

On a fundamental level, this field relates the properties and performance of a material to its atomic-scale structure and the different phases it can go through. The major factors that determine the structure and properties of a material are the nature of its constituent chemical elements and the way in which the material was processed into its final form. These factors, related through the laws of thermodynamics, govern the material's microstructure, and thus its properties.

An old adage in materials science says: "materials are like people; it is the defects that make them interesting". Given the limits of today's technology, that is good, because the manufacture of a perfect crystal of a material is physically impossible. Instead, materials scientists manipulate a material's defects to create materials with the desired properties. On an atomic scale, the defects in a crystal could mean that atoms of one element may be missing or replaced by atoms of other elements.

Not all materials have a regular crystalline structure. Glasses and some ceramics—unlike many natural materials— are amorphous, that is, they do not possess any long-range order in their atomic arrangements. Engineering these materials is much more difficult than engineering crystalline materials. Polymers may exhibit varying degrees of crystallinity, and studying them requires a combination of elements of chemical and statistical thermodynamics to give thermodynamic (rather than mechanical) descriptions of physical properties.

Subfields of Materials Science

- Nanotechnology: As commonly understood, nanotechnology is the field of applied science and technology concerned with the formation, study, and control of materials having a width ranging from less than 1 nanometer (10^{-9} meter) to 100 nanometers. These materials are generally engineered on a molecular scale. On a more rigorous level, nanoscience involves the study of materials whose defining properties are present only at the nanoscale.
- Crystallography: This is the study of the arrangement of atoms in a crystalline solid and the relationship between the crystalline structures and their physical properties. It includes the determination of defects associated with crystal structures.
- Materials characterization: Information needed for understanding and defining the properties of materials is acquired through such techniques as diffraction of X-rays, electrons, or neutrons, and various forms of spectroscopy, chromatography, thermal analysis, or electron microscopy.
- Metallurgy: This involves the study of metals and their alloys, including their extraction, microstructure, and processing.
- Tribology: This is the study of the wear of materials due to friction and other factors.
- Surface science: It involves study of the structures and interactions occurring at the interfaces of solids and gases, solids and liquids, and solids and solids.
- Glass science: It involves the study of non-crystalline materials, including inorganic glasses, vitreous metals, and non-oxide glasses.

Some practitioners consider rheology a subfield of materials science, because it can cover any material that flows. Modern rheology, however, typically deals with non-Newtonian fluid dynamics, so it is often considered a subfield of continuum mechanics.

- Thermodynamics statistical mechanics, chemical kinetics, and physical chemistry: To understand phase stability and physical and chemical transformations.
- Chemical bonding: To understand the bonds between atoms of the material.
- Mechanics of materials: To understand the mechanical properties of materials and their structural applications.
- Solid-state physics and quantum mechanics: To understand the electronic, thermal, magnetic, chemical, structural, and optical properties of materials.

- Solid-state chemistry and polymer science: To understand the properties of polymers (including plastics), colloids, ceramics, and liquid crystals.
- Biology: For the integration of materials into biological systems.
- Continuum mechanics and statistics: For the study of fluid flows and ensemble systems.
- Diffraction and wave mechanics: For the characterization of materials.

2

Classification of Materials

Materials are generally classified on the basis of their properties into metals, nonmetals, metalloids, alloys, composites, polymers and ceramics. Metals are further divided into ferrous metals and non-ferrous metals. The topics elaborated in this chapter will help in gaining a better perspective about these types of materials.

Metals

Metals are the most commonly used class of engineering material. Metals are employed for various engineering purpose and to requirements, such as structural members, pipes, tanks, and building up of engineering machines. Out of all metals, Iron is the most popular metal in the field of engineering. All the metals have a crystalline structure.

Properties of Metals

Physical Properties of Metals

Metals are lustrous, malleable, ductile, good conductors of heat and electricity. Other properties include:

- State: Metals are solids at room temperature with the exception of mercury, which is liquid at room temperature (Gallium is liquid on hot days).
- Luster: Metals have the quality of reflecting light from their surface and can be polished e.g., gold, silver and copper.
- Malleability: Metals have the ability to withstand hammering and can be made into thin sheets known as foils. For example, a sugar cube sized chunk of gold can be pounded into a thin sheet that will cover a football field.
- Ductility: Metals can be drawn into wires. For example, 100 g of silver can be drawn into a thin wire about 200 meters long.
- Hardness: All metals are hard except sodium and potassium, which are soft and can be cut with a knife.
- Valency: Metals typically have 1 to 3 electrons in the outermost shell of their atoms.
- Conduction: Metals are good conductors because they have free electrons. Silver and copper are the two best conductors of heat and electricity. Lead is the poorest conductor of heat. Bismuth, mercury and iron are also poor conductors.

- Density: Metals have high density and are very heavy. Iridium and osmium have the highest densities whereas lithium has the lowest density.
- Melting and Boiling Points: Metals have high melting and boiling points. Tungsten has the highest melting and boiling points whereas mercury has the lowest. Sodium and potassium also have low melting points.

Chemical Properties of Metals

Metals are electropositive elements that generally form basic or amphoteric oxides with oxygen. Other chemical properties include:

- Electropositive Character: Metals tend to have low ionization energies, and typically lose electrons (i.e. are oxidized) when they undergo chemical reactions They normally do not accept electrons. For example:
    - Alkali metals are always 1^+ (lose the electron in *s* subshell).
    - Alkaline earth metals are always 2^+ (lose both electrons in *s* subshell).
    - Transition metal ions do not follow an obvious pattern, 2^+ is common (lose both electrons in *s* subshell), and 1^+ and 3^+ are also observed.

$Na^o \rightarrow Na^+ + e^-$

$Mg^o \rightarrow Mg^{2+} + 2e^-$

$Al^o \rightarrow Al^{3+} + 3e^-$

Compounds of metals with non-metals tend to be ionic in nature. Most metal oxides are basic oxides and dissolve in water to form metal hydroxides:

$Na_2O\ (s) + H_2O\ (l) \rightarrow 2\ NaOH(aq)$

$CaO\ (s) + H_2O\ (l) \rightarrow Ca(OH)_2\ (aq)$

Metal oxides exhibit their *basic* chemical nature by reacting with *acids* to form metal *salts* and water:

$MgO(s) + HCl\ (aq) \rightarrow MgCl_2\ (aq) + H_2O(l)$

$NiO(s) + H_2SO_4\ (aq) \rightarrow NiSO_4(aq) + H_2O(l)$

Ferrous Metals

Ferrous metals are those which contain iron as their main constituent or base metal. Iron (Fe) is the principal element of Ferrous Metals.

The ferrous metals include all the types of Iron, Steel, and their alloys. In the present, their role in the engineering industries can be easily described as "most dominating."

In all the jobs ranging from the manufacture of a primitive type of agricultural implements to advanced types of Air Crafts, ferrous metals and their alloys occupy a prominent position.

In the automotive, building and bridge construction, railways, light, and heavy machinery, shipping, and transportation.

And in any other field of engineering activity, it may not be possible to move ahead without metals. This is explained by a number of reasons:

1. The wide abundance of Iron ore in almost all parts of the world.
2. The economic extraction of Iron from its ore.
3. The flexibility that can be induced in the mechanical properties of Iron by combining it with other metals and by heat treatment and such other methods.

These facts explain the reason that for the considerable time, the annual global production of Ferrous Metals has been far in excess than the combined production of all Non-Ferrous Metals produced in all the countries of the world.

Ferrous Metals List, their Types and Properties

Pig Iron

Pig Iron is the first or basic form in which Iron is prepared as the metal from its ores.

It is, therefore, impure and crude and requires subsequent processing to develop Cast, Wrought Iron, and Steel, which are the common Ferrous Metals used in construction and industries.

Following are a few types of Pig Iron distinguished on the basis of their properties:

1. Grey Pig Iron: It is also called foundry pig. It is soft in character and rich in carbon. It is produced when the raw material is burnt at a very high temperature.
2. White Pig Iron: It is also called forge Pig Iron, as it is hard and brittle and can be converted only by using pressure. This type contains sulfur as the main impurity and hence is considered inferior in grade.
3. Bessemer Pig Iron: It is specially used in the manufacture of Steel in the Bessemer process, because of its freedom from sulfur and phosphorous.

Cast Iron

Cast Iron is derived from the Pig Iron. Pig Iron, because of its impurities, is weak and hence very difficult to shape into various forms. Therefore, Pig Iron is remolded in a furnace and cast or poured into molds of the desired shape to get the Iron known as Cast Iron.

Manufacturing of Cast Iron

Cast Iron is manufactured in a furnace known as cupola furnace. Cupola furnace is cylindrical in shape with an almost uniform diameter (about 1 meter) and a height of about 5 to 6 meters. The cylinder has an inner lining of Refractory Bricks which is provided with Tuyers near the bottom for injecting the supply of air blast.

The outer shell is made of Steel. The top of the furnace remains open. The cupola furnace contains a charging door, a platform and tapping holes for the purpose of blasting the air.

The working of the furnace or manufacturing of Cast Iron is as follows:

- The mixture of Pig Iron, coke, and limestone, known as a charge, is prepared in correct proportions. Coke is used as fuel and limestone as a fluxing material.
- Flux is the material which easily fuses and mixes with impurities to form a slag, which can be taken through the top of the furnace.
- The molten metal in almost pure form is collected at the bottom of the furnace along with the floating slag. The slag is removed intermittently through the hole and molten metal taken out through its tap hole.
- The molten metals taken out are fed into the molds prepared to get desired shapes. The Iron obtained is known as Cast Iron, and the shapes obtained are called Cast Iron castings.
- The Cast Iron thus obtained, has 1.7 to 4% of carbon with small impurities of manganese, phosphorus, silicon, and sulfur.

General Properties of Cast Iron

1. It is hard and brittle.
2. It can be hardened but cannot be tempered.
3. It cannot be forged, welded, or riveted.
4. It cannot be magnetized.
5. It does not rust easily.
6. It becomes soft in saltwater.
7. It shrinks easily.

8. It is weak in tension and strong in compression.
9. It is not suitable for shock and impact loads.
10. It has a specific gravity of 7.5, and it has a melting point of 1250 °C.
11. It has good machinability property.

Uses of Cast Iron

1. Used for water and gas pipes.
2. Used in manufacturing of cisterns, sewer and drain pipes, manhole covers and fittings.
3. Used for rail chairs, carriage wheels, etc.
4. Used in compression members, bed plates, etc.
5. Used for making parts, bodies, and beds of machines.

Types of Cast Iron

Following are the common types of Cast Iron used in engineering materials.

Grey Cast Iron

Grey Cast Iron is produced by melting the foundry Pig Iron or grey Pig Iron in the cupola furnace. It shows grey color when fractured. The grey color is due to the presence of free graphite.

The usual composition of Grey Cast Iron is:

- Iron – 92 percent.
- Carbon – 3 to 3.5 percent (as graphite).
- Carbon – 0.5 percent (combined).
- Silicon – 1 to 2.75 percent.

- Manganese – 1 percent.
- Phosphorus – 0.5 to 1 percent.
- Sulfur – 0.02 to 0.15 percent.

It is soft and ductile. It is commonly used in castings, dies, molds, machine frames, and pipes, etc.

Properties of Grey Cast Iron

1. It is gray in color.
2. It is soft compared to other Cast Irons and has good machinability.
3. It has high compressive strength.
4. It is poor in tensile strength and impact strength with almost no ductility.

Uses of Grey Cast Iron

1. It is used for machine tool bodies.
2. It is used for pipe and pipe fittings.
3. It is used for automobile cylinder blocks, heads, housings, etc.
4. It is used in flywheels and agricultural implements.

White Cast Iron

It is white in color as there is no free graphite. In white Cast Iron, the entire carbon is in combined state. i.e., in the form of cementite, also known as Iron carbide. White Cast Iron is produced by melting the low phosphorus Pig Iron along with the Steel scraps in a cupola furnace.

The Cast Iron thus obtained is in a molten state, and it is chilled, i.e., it is cooled rapidly. Therefore, it is also known as chilled Cast Iron. Due to rapid cooling, its outer surface becomes harder while its interior remains softer.

It has a shiny white color and a bright metallic white luster. It is very strong, hard, and resistant to Wear and Tear. And quite brittle as well.

Its usual composition is as follows:

- Iron – 94 percent.
- Carbon – 1.75 to 2.3 percent.
- Silicon – 0.85 to 1.2 percent.
- Manganese – 0.1 to 0.4 percent.
- Phosphorous – 0.05 to 0.2 percent.
- Sulfur – 0.12 to 0.35 percent.

Properties of White Cast Iron

1. It is white in color.
2. It is hardest of all Cast Irons and is wear-resistant.
3. It is brittle.
4. It has high tensile strength but weak compressive strength.
5. It is not easily machinable due to its hardness, hence, require special tools for machining.

Uses of White Cast Iron

It is used for car wheels, rollers for crushing grains, crusher jaw plates, etc.

Malleable Cast Iron

Malleable Cast Iron is the annealed White Cast Iron, i.e., white Cast Iron is heated slowly up to 900 to 950 °C temperature for several days and then cooled at a slower rate. The Iron obtained is malleable as well as ductile. It possesses useful properties of both Cast Iron and mild Steels.

The tensile strength of malleable Cast Iron is higher than that of grey Cast Iron and has good machinability. It is used for hubs of wagon wheels, railway rolling stock, brake supports, parts of agricultural machinery, pipe fittings, door hinges, locks, etc.

Uses of Malleable Cast Iron

1. It is used in Automobile industries for making rear-axle housing, steering-gear housing, hubs, and pedals, etc.
2. It is used in Railways equipment of a great variety.
3. It is used in Agricultural machinery making and carpentry tools.

Ductile Cast Iron

It is a modified Grey Cast Iron. It is also called as nodular or spheroid graphite Cast Iron or high strength Cast Iron.

Ductile Cast Iron is produced by adding any one of the elements of magnesium, calcium, cerium, bismuth, zinc, cadmium, titanium, and boron into the molten Grey Cast Iron.

Composition of ductile Cast Iron is:

- Carbon – 3.2 to 4.5%.
- Silicon – 1 to 4%.
- Magnesium – 0.1% to 0.8%.
- Phosphorus – 0.1%.
- Iron – the rest % is Iron.

Properties of Ductile Cast Iron

1. It has high fluidity.
2. It has high tensile strength, toughness, and wears resistance.

Uses of Ductile Cast Iron

1. It is used for castings where shock and impact loads are operating.
2. It is used in rolls for rolling mills, hydraulic cylinders, cylinder heads, etc.

Alloy Cast Iron

Alloy Cast Irons are also called as plain Cast Irons. Alloyed Cast Irons are produced by adding the alloying elements to the Pig Iron in the cupola furnace itself or adding the same into the molten Cast Iron taken out of the furnace.

Better results are obtained when Pig Iron with alloying elements are heated in an electric furnace or air furnace. The alloying elements used are; nickel, chromium, molybdenum, and vanadium.

The usual alloying elements are *Nickel* and *Chromium*. *Nickel* is added to affect an increase in the hardness and resistance to wear. Chromium makes the Cast Iron extremely tough, strong and ductile.

The alloy Cast Irons are used for gears, sprockets, wheels, I.C engine cylinders, piston, piston rings, crankcases, crankshafts, camshaft, brake drums and shoes, parts of grinding machines, etc.

Wrought Iron

It is the purest form of Iron, containing all impurities below a limit of 0.5 percent. And carbon is included in these impurities, its proportion being generally less than 0.12 percent.

Besides, Wrought Iron always contains a small proportion of slag in the silicate component. The source material for the manufacture of Wrought Iron is PIG IRON. There are two processes for the manufacture of Wrought Iron:

1. Puddling Process
2. Aston Process

Properties and uses of Wrought Iron

Its most important properties are mentioned below:

1. Strength: It has a tensile strength varying between 2500 to 4000 kg/cm^2. The strength will be greater in the longitudinal direction.
2. The ultimate compressive strength ranges between 1500-2000 kg/cm^2.
3. Physical: Wrought Iron is bluish in color, has a silky luster and fibrous structure. It is malleable, ductile, tough, and resistant to corrosion.
4. Density: The metal has a density of 7.8 gm/cm^3 and a melting point of 1500 centigrade.

Wrought Iron shows good resistance to fatigue and sudden shock. Moreover, it can be welded with ease.

Because of the above set of properties, Wrought Iron is extensively used as a material for making plates, sheets, pipes, tubes, etc.

It is also used in buildings, railways, and marine industries.

Steel

Steel is an alloy of Iron and carbon, where the carbon content is less than 1.7%. If the carbon content in Steel exceeds 1.7%, it does not combine with the Iron, but it is present as free graphite. Besides carbon, many other metals may also be present in addition to Iron, giving rise to great varieties of Steel.

On the basis of the presence of free graphite, differentiation of Steel and Cast Iron can be made. If there is a free graphite present, it is a Cast Iron, otherwise a Steel. The hardness and toughness of Steel increase with the increase in carbon content up to 1.7%.

On the other hand, with the decrease in the carbon content (lower than 0.1%), the material would resemble more to Wrought Iron or pure Iron. The best thing about Steel is that it has both properties of Cast and Wrought Iron – Compressive Strength of Cast Iron and Tensile Strength of Wrought Iron. Due to these properties, Steel is used as a structural material in all types of situations.

Steel is a versatile material of modern age. Its properties can be varied over a wide range by varying its composition and by subjecting it to various mechanical and heat treatment processes. As we noted earlier, Cast Iron is better in resisting compressive stresses, while Wrought Iron is suited to tensile stresses. Steel is superior in resisting both compressive and tensile stresses.

Hence, Steel finds most of the applications for all purposes in places of Cast Iron and Wrought Iron. The elements of the composition of Steel apart from carbon are sulfur, silicon, phosphorus, manganese, etc.

Manufacturing of Steel

Steel is manufactured using one of the following processes:

1. Bessemer process
2. Open hearth process
3. L-D process
4. Crucible process
5. Electric process
6. Duplex process
7. Cementation process

Classification of Steels

Steels can be classified in many ways such as on the basis of the methods used in their manufacture, on the carbon content, or according to their use.

It can also be classified on the basis of Steel casting. It is, however, the classification on the basis of their chemical composition is commonly adopted. Following are the major groups of Steels.

Plain Carbon Steels

This is the first major group of Steels. Carbon is the only controlling component in them besides Iron. They are further subdivided into 3 subcategories.

- Low Carbon Steels (C=0.05-0.25%)
- Medium Carbon Steels (C=0.25-0.50%)
- High Carbon Steels (C=0.50-1.50%)

Alloy Steels

An alloy is the purposeful mixture of two or more elements of which one being in largest proportion is called as base metal, and other elements are called as alloying elements. However, the Steel, containing Iron and carbon, is not referred to as an alloy. But if elements are added other than Iron and carbon, the Steel is known as alloy Steel.

Thus all the Steels, in addition to Iron and carbon-containing other elements such as; nickel, chromium, manganese, silicon, vanadium, molybdenum, tungsten, sulfur, phosphorus, etc., are called as alloy Steels. The purpose of alloying is to improve the properties of Steel, like imparting the fine grain size, to improve the hardness, toughness, strength, corrosion resistance, etc.

These are Steels made with the addition of a definite proportion of a selected element or elements in addition to carbon at the manufacturing stage.

Benefits of alloying are as follow:

- The tensile strength of the Steel may be increased without affecting its workability.

- The resistance against very high temperature, abrasion and corrosion may be improved considerably.
- The electrical, magnetic, and thermal properties may be modified in the desired direction.

They can be sub-divided into 2 types on the basis of the proportion of alloying elements, which are the following:

- Low Alloy Steels
- High Alloy Steels

Nonferrous Metals

Nonferrous metals do not contain any iron or metals that have an insignificant amount used as an alloy. Nonferrous metals are nonmagnetic. Standard nonferrous metals are aluminum, copper, lead, nickel, tin, and zinc and their alloys. Nonferrous metals have lower strength at high temperatures and more shrinkage than ferrous metals.

Nonferrous Metals (Top left: Tungsten, Top right: Aluminum, Bottom Left: Brass; Bottom Right: Copper).

Advantages

Benefits of nonferrous metals are:

- Attractive colors
- Low density
- Excellent electrical conductivity
- Good thermal conductivity

- Easy to roll, forge, weld, cast, machine and fabricate
- High corrosion resistance

Uses and Properties of Nonferrous Metals

Welding Nonferrous Metals

There are many welding applications of nonferrous metals. However, the weld ability varies and is different than when you are welding steel. Before welding review the procedures that are unique to each type of metal.

Aluminum (AI)

Aluminum has the symbol AI and is a chemical element in the boron group. It has the atomic number 13. It is malleable, gray-white-silver in color, nonmagnetic, soft and ductile metal. It is easy to work with and has an attractive appearance.

How it is Produced?

Aluminum is produced by electrical processes from a clayey mineral known as bauxite. Commercial aluminum alloys are classified into two groups:

- Wrought alloys: Includes those alloys designed for mill products whose final physical forms are obtained by working the metal mechanically.
- Cast alloys: Includes those alloys whose final shapes are obtained by allowing the molten metal to solidify in a mold.

Welding an Aluminum Boat For High Yield Strength.

Uses

- Household utensils and appliances
- Alloying agent and deoxidizer in steel manufacturing

- Castings
- Pistons
- Torque converter pump housings
- Aircraft structures
- Kitchen utensils
- Railways cars
- Transmission lines

Capabilities

Aluminum can be:

- Cast
- Forged
- Machined
- Formed
- Welded

Limitations

Avoid direct contact of aluminum with copper alloys and copper. Use aluminum in low-temperature applications.

Properties

Aluminum is lightweight, soft and a low strength metal that needs alloys added to increase strength. Aluminum can be easily cast, forged, machined, formed, and welded.

Except when alloyed with specific elements, it is only for low-temperature applications.

- Corrosion Resistant
- High strength/weight ratio at room temperature
- Corrosion resistant
- High electrical conductivity (Volume for volume, 60% of copper)
- Excellent thermal conductor
- Lightweight
- High ductility

- Needs to be frequently annealed if cold worked
- Hardens after cold
- Malleable

Aluminum Properties

- Hardness: Pure aluminum has a Brinell hardness number of 17 to 27.
- Strength: Aluminum has a tensile strength of 6000 to 16,000 psi (41,370 to 110,320 kPa).
- Gravity: specific gravity of 2.7.
- Melting Point: The aluminum melting point is 1220 °F (660 °C).
- Density 20 °C: 2.702 g/cm^3.

Aluminum Alloy Properties

- Hardness: Aluminum alloys have a Brinell hardness number of 100 to 130.
- Strength: tensile strength of 30,000 to 75,000 psi (206,850 to 517,125 kPa).

Welding Aluminum

Aluminum can be welded using mechanical fastening, adhesive bonding, soldering, brazing, and welding (most common.)

Welding Processes that are used with aluminum:

- Arc Welding (most common)
    - GTAW (gas tungsten arc welding)
    - GMAW (gas shielded metal arc welding)
- Stud Welding
- Electron Beam Welding
- Laser Beam Welding
- Oxyfuel Gas Welding
- Resistance Welding
- Solid-state Welding

Identification Tests

- Appearance Test: Aluminum is silver to light gray and when polished is very bright. It gets dull when it becomes oxidized. It is also light weight.

- Rolled sheet aluminum is typically usually pure metal.
- Aluminum alloy castings are usually made with copper, zinc, silicon, and sometimes magnesium and iron.
- Wrought aluminum alloys may contain manganese, magnesium, silicon or chromium.
- Magnesium and aluminum are similar in appearance. Magnesium is distinguished from aluminum by applying on the surface a drop of silver nitrate solution. The silver nitrate does not react to the aluminum but leaves a black silver deposit on the magnesium.

- Fracture Test: A fracture in an aluminum casting shines a bright crystalline structure.
- Spark Test: No sparks are given off from aluminum.
- Torch Test: Before melting aluminum does not turn red. It holds its' shape just before becoming molten. Once it does it then suddenly collapses (hot shorts). On the molten surface a heavy film of white oxide forms.

Aluminum Alloys

One or more metals are alloyed with Aluminum including nickel, silicon, manganese, magnesium, and copper. Aluminum is strengthened and hardened with small amounts of alloyed metals.

Aluminum vs. Copper

Aluminum has replaced copper due to its malleability and ductility, particularly in appliances and for electrical transmission. When conducting heat and electricity, aluminum is not as good as copper.

Antimony (Sb)

Antimony is made from stibnite (also called antimonite.) It is a by-product of refining zinc, copper silver or lead. Alloy in heavy metals such as copper, tin, and lead.

Uses and Properties

- Color: Silver white
- Hard
- Brittle (can be powdered)
- Very crystalline
- Melting Point: 1,167 °F (630.6 °C)
- Boiling Point: 2,888 °F (1,587 °C)

Brass and Bronze

Brass and bronze are copper alloys. As a copper alloy, they are corrosion resistant and good conductors of electricity and heat.

Brass

All Brass is an alloy of copper and zinc (60 to 68 percent copper and 32 to 40 percent zinc). It has high heat conductivity and a low melting point.

True Brass is an alloy of copper and zinc. It sometimes has additional alloys for specific properties. Several grades of strips and sheets are available.

All Brass Is An Alloy of Copper and Zinc.

Types of Brass

There are several types of brass:

- Naval

- Red
- Admiralty
- Yellow
- Cartridge
- Muntz metal leaded
- Admiralty
- Nickel brass
- Commercial:
    - Afta-Betabrass: Contains 36% to 45% Zinc with the rest comprised of copper for hot work.
    - Alfa Brass: Contains up to 36% Zinc with the remaining part copper for cold working.

All types vary in levels of zinc and copper. Each can be alloyed with other elements such as lead, tin, manganese, or iron. They have excellent machinability and can be welded.

Properties

- Ductility and tensile strength increases with the amount of Zinc (up to 30%).
- Different types can be cast, hot forged, cold forged, cold rolled into sheets, drawn into wires and extrusions for cross-section bars.
- Beyond 30% Zinc, the tensile strength increases up to 45% of Zinc, but the ductility significantly drops. β-phase is less ductile than α-phase, but it is stronger and harder.
- Melting point: between 840°C to 960°C depending on the type of brass.
- Machinability: Improves if 1% to 2% zinc is added.
- Easily soldered.
- Corrosion resistant.

Uses

- Bushes
- Bearings
- Utensils
- Pump linings
- Hydraulic fittings

Brass vs. Copper

Brass has lower electrical conductivity and has a lower thermal conductivity than copper. It is stronger than copper.

Welding Brass

The Zinc content in brass makes brass with a high concentration of Zinc harder to weld. The vaporization of Zinc during the welding process results in blowholes and incomplete fusion. It also degrades weld ability and workability.

For this reason, choose a brass with a low concentration of Zinc. Processes for welding Brass are:

- GTAW (recommended to minimize zinc fuming)
- GMAW

Bronze

The name Bronze is currently applied to alloys of copper, tin and any copper-based alloy that looks like bronze.

It may contain zinc, lead, nickel, phosphorus or manganese. It is corrosion resistant, has good machinability, high strength and rust resistance.

Bronze is a Tin and Copper Alloy.

Types of Bronze

Welding

Bronze can be welded with differences between the different bronze alloys.

Copper Composition

To be useful, Bronze contains 75% to 95% copper with the rest of the metal made up of tin. Higher levels of tin content result in higher strength and greater corrosion resistance (called hot working bronze).

Uses

- Alloy can be cold rolled into sheets, rods, and wire

- Drawn and stamped products
- Rods
- Sheets
- Utensils
- Bushes
- Bearings
- Hydraulic fittings

Brass vs. Bronze

Bronze has more corrosion resistance and better mechanical properties than brass.

Brass and Bronze Tests

- Appearance Tests: The color of polished bronze and brass varies with a shift from a red copper color to yellow brass. They oxidize to various shades of yellow, green, and brown.
- Fracture Test: The surface of fractured brass or bronze ranges from smooth to crystalline, depending upon composition and method of preparation (cast, rolled, or forged).
- Spark Test: Sparks are not given off by brass and bronze.
- Torch Test: Brass contains zinc, which when melted gives off white fumes. Conversely, the metal Bronze has tin. A small amount of tin makes the alloy flow like water. Due to the small amount of zinc or tin that is usually present, bronze may fume slightly, but never as much as brass.

Aluminum Bronze Tests

- Appearance: When polished, aluminum bronze appears a darker yellow than brass.
- Fracture: When fractured, Aluminum Bronze has a smooth surface.
- Spark Test: No sparks are given off by aluminum bronze.
- Torch Test: It is tough to weld aluminum bronze. A difficult to remove scum covers the surface, and that mixes with the metal.

Cadmium (CD)

Cadmium is a by-product of lead and zinc. It is extremely toxic and frequently found in manufacturing.

Uses and Properties

- Color: White metal with a blue hue
- Can be polished
- Specific gravity is 8.67
- Melting point at 321 °C
- Malleable
- Ductile
- Quickly drawn into wires and rolled

It is softer than zinc and harder than tin.

Chromium (Cr)

Chromium is an alloying agent used in steel, cast iron, and nonferrous alloys of nickel, copper, aluminum, and cobalt. It is hard, brittle, corrosion-resistant, can be forged, machined, welded and electroplated. Chromium is not resistant to hydrochloric acid and cannot be used in its pure state because of its difficulty to work.

Chromium is Brittle and Hard Nonferrous Metal with a High Melting Point.

Uses

One of the most widely used alloys is chromium. In steel and cast iron it is used as an alloying agent (0.25 to 0.35 percent) and the nonferrous alloys of cobalt, copper, nickel, and aluminum. It is also used in powder metallurgy, electroplating for wear and appearance, and to make stainless steel and mirrors.

Capabilities

Chromium alloys can be forged, machined and welded. Chromium is never used in its pure state.

Limitations

Chromium is brittle, making it difficult to work with in a pure state. It is not resistant to hydrochloric acid.

Properties

- Specific gravity of 7.19.
- Melting point of 3300 °F (1816 °C).
- Brinell hardness number of 110 to 170.
- Resistant to acids other than hydrochloric, and is heat, wear, and corrosion resistant.

Cobalt (Co)

Cobalt is a hard, white metal similar to nickel in appearance, but has a slightly bluish cast.

Cobalt Alloy Engine Part (right).

Uses

Cobalt does not have many uses in a metallic form. When combined with other elements, it is used for hard-facing materials. As an alloy Cobalt is mainly used in:

- Permanent materials.

- Soft magnetic materials.
- High-speed cutters and tool bits.
- High-temperature, creep-resisting alloys.
- Cemented carbide tools, bits, and cutters.
- Used in making blue ceramic glazes and insoluble paint pigments.

Capabilities

Cobalt can be machined (limited) welded, and cold-drawn.

Limitations

Cemented carbide cutters are used to machine Cobalt. Cracking can occur when welding high carbon cobalt steel.

Properties

- Tensile strength of 34,000 psi (234,430 kPa).
- Brinell hardness number of 125.
- Specific gravity of 8.9.
- Melting point of 2720 °F (1493 °C).

Cobalt Alloy (Stellite 21) Properties

- Tensile strength of 101,000 psi (696,395 kPa).
- Heat and corrosion resistant.

Copper (Cu)

Copper Alloys Are The Best Conductors of Electricity and Heat.

Copper is one of the most popular commercial metals. It is used with many alloys. It is frequently used as a protective coating for rods and sheets to make ball floats, containers, and soldering coppers.

Though it is very soft, it is tough to machine due to its high ductility and malleability. Copper is a reddish metal and has high electrical and heat conductivity. It is used as a primary element in hundreds of alloys. Commercially pure copper is not suitable for welding. Gas welding is the preferred process for joining copper and copper alloys and copper.

Types of Copper

- Beryllium copper (1.50 to 2.75% beryllium): When hardened it gains in tensile strength but loses ductility. It is ductile when soft.
- Blister Copper: The crude form of copper is called Blister copper (68% pure).
- Pure Copper: With electrolytic refining, highly pure (99.9%) copper is remelted and cast into required shapes.
- Nickel-copper (Cupronickel): Contains either 10, 20, or 30 percent nickel. Nickel alloys have moderately high to high tensile strength, which increases with the nickel content. They are somewhat hard, quite tough, and ductile. They are very resistant to the erosive and corrosive effects of high-velocity sea water, stress corrosion, and corrosion fatigue. Some alloys are used for salt water piping systems; other sheet forms are used to construct small storage tanks and hot water reservoirs. To avoid blowholes when welding use the filler metals of ERCuNi for GTAW and GMAW. If welding with SMAW use the filler metal ECuNi.
- Nickel Silver (Copper-Zinc Alloys): Nickel is added to copper zinc alloys (brasses) to lighten their color; the resultant alloys are called nickel silver. These alloys are of two general types:
    - Contains 65 percent or more nickel and copper combined. Can be cold worked by such operations as deep drawing, stamping, and spinning.
    - Contains 55 to 60 percent copper and nickel combined. The much harder end is not processed by any of the cold working methods.

Copper Welding

Copper and copper alloys are joined using soldering, brazing, and welding.

Welding processes used are:

- Arc Welding
    - GTAW (Gas Tungsten Arc Welding)
    - GMAW (Gas Metal Arc Welding)
    - SMAW (Shielded Metal Arc Welding, but only for non-critical applications)

- Electron Beam Welding
- Flash Welding
- Friction Welding
- Laser Welding
- Oxyfuel Gas Welding
- Pressure Welding
- Ultrasonic Welding
- Resistance Welding

Copper Uses

Commercially pure copper is primarily utilized in the electrical industry where it is made into electrical conductors such as wire.

Non-ferrous alloys such as bronze, brass, and Monel are manufactured using copper.

Other Uses

- Sheet roofing
- Cartridge cases
- Bushings
- Statues
- Bearings
- Wire

Capabilities

Copper is cast, cold worked and forged. Machinability is fair, but copper and copper alloys can be welded with some limitations.

Limitations

There are welding limitations for electrolytic tough pitch copper. Pure copper is tough to machine due to ductility and is not suitable for welding.

Properties

- Nonmagnetic (Pure copper)
- Hardness: Brinell hardness number of 60 to 110
- Strength: Copper has a tensile strength of 32,000 to 60,000 psi (220,640 to 413,700 kPa)

- Copper alloys:
  - Tensile strength: 50,000 to 90,000 psi (344,750 to 620,550 kPa)
  - Brinell hardness: 100 to 185
- Gravity: It has a specific gravity of 8.9
- Melting Point: melting point of 1980 °F (1082 °C)
- Corrosion resistant

Copper Tests

- Appearance: When polished, copper is red, it oxidizes to a green shade.
- Fracture Test: When fractured, copper as a smooth surface and does not have a crystalline appearance.
- Spark Test: No sparks are given off by Copper.
- Torch Test: Because copper conducts heat rapidly, a larger flame is required to fuse copper than is needed for the same quantity of steel. Copper solidifies instantly and melts suddenly. Copper alloy, which contains small amounts of other metals melts more easily and hardens more slowly than pure copper.

Lead (Pb)

Lead is a heavy metal (heaviest of all common metals) that is soft and malleable. It has a low melting point, creep strength and tensile strength. Lead extraction is from galena ore.

Color

The surface is bluish grayish in color, but after scratching or scraping it, the actual color of the metal appears white. When exposed to air it gets a very dull luster.

Capabilities

Welding and Example of a Lead Roof.

Lead can be machined, welded, cast, and cold worked. It is moisture, corrosion, atmosphere, and water resistant, and is resistant to many acids.

The correct welding rod selection is necessary due to a low melting point.

Limitations

Lead has low strength with heavy weight. Lead dust and fumes are very poisonous.

Properties

- Pure Lead: Pure lead has a tensile strength of 2500 to 3000 psi (17,237.5 to 20,685 kPa); specific gravity of 11.3; and a melting point of 620 °F (327 °C).
- Alloy Lead: Alloy lead B32-467 has a tensile strength of 5800 psi (39,991 kPa). It is alloyed with steel and brass to improve machinability. It is also alloyed in plumbers and other solders.

General Properties

- Low electrical conductivity
- Self-lubricating
- Malleable; can be easily shaped and worked
- Corrosion resistant
- Chemical resistant (even many acids)
- Soft
- Heavy

Uses

Lead has more fields of application than any other metal.

- Electrical equipment (main use)
    - Lead-coated power
    - Electrical cables
    - Telephone cables
    - Storage batteries
- Building construction
    - Pipe and sheet form
    - Tanks

- Roof coverings
- In solder

Uses of Lead-zinc Alloys

- Utilized in the manufacture of:
  - Lead weights
  - Bearings
  - Gaskets
  - Seals
  - Bullets
  - Shot
- Chemical compounds
  - Lead carbonate (paint pigment)
  - Tetraethyl lead (antiknock gasoline)
- X-ray protection (radiation shields)

Caution

Lead dust, fumes, and vapors are poisonous. Use care and protective safety equipment when welding lead.

Magnesium (Mg)

Magnesium has many useful properties. It is a very light metal (lighter than aluminum) making it easy to cast and machine. It is white in color, has a low melting point, excellent machinability, Magnesium has some moderate levels of resistance to atmospheric exposure, many chemicals such as alkalies, chromic and hydrofluoric acids, hydrocarbons, and most alcohols, phenols, esters, and oils. It is nonmagnetic. Galvanic corrosion is an important factor in any assembly with magnesium.

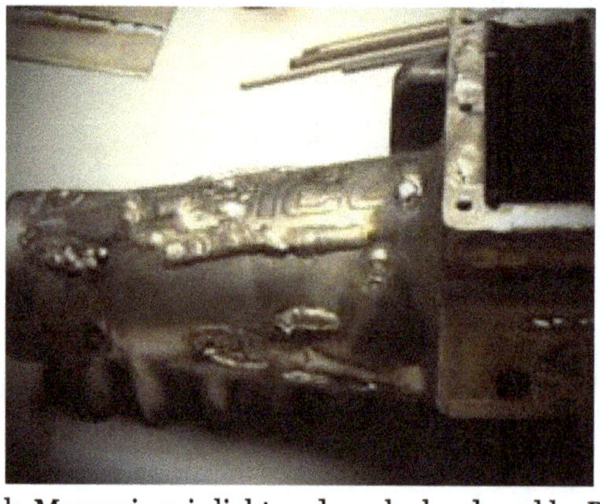

Magnesium Engine Block. Magnesium is light and can be hardened by Precipitation Hardening.

It is nonmagnetic. Galvanic corrosion is an important factor in any assembly with magnesium. Principal magnesium ores are dolomite, carnallite and magnesite. It is extracted using an electrolytic process.

Magnesium is moderately resistant to:

- Atmospheric exposure
- Chemicals such as:
    - Alkalies
    - Hydrofluoric acids and chromic acids
    - Hydrocarbons
    - Most alcohols
    - Phenols
    - Esters
    - Oils

Welding

Magnesium is weldable by either the arc or gas processes. It requires the use of a gaseous shield. It is welded on AC, but you want to weld them with the right filler metal.

Capabilities

Magnesium can be forged, cast, welded, and machined.

Limitations of Magnesium

Magnesium in fine chip form will ignite at low temperatures (800 to 1200 °F (427 to 649 °C)). The flame can be mothered with suitable materials such as carbon dioxide (CO_2), foam, and sand.

Properties

Magnesium is the lightest metal. It weights two-thirds the weight of aluminum. The tensile strength of cast magnesium metal equals cast aluminum (90 MPa.) The tensile strength of rolled annealed magnesium is equal to quality cast iron.

Magnesium can be drawn forged, accurately machined and easily formed.

Powdered magnesium easily burns (use with fire protection and caution).

- Tensile Strength: Pure magnesium has tensile strength of 12,000 psi (82,740 kPa) (cast) and tensile strength of 37,000 psi (255,115 kPa) (rolled).
- Hardness: Brinell hardness number of 30 (cast) and 50 (rolled).

- Gravity: Specific gravity of 1.7.
- Melting Point: Melting point of 1202 °F (650 °C).

Magnesium Alloy Properties

- Hardness: Brinell hardness number of 72 (hard) and 50 (forged).
- Strength: Tensile strength of 42,000 psi (289,590 kPa) (hard) and 32,000 psi (220,640 kPa) (forged).

Uses

- Deoxidizer for bronze, brass silver and nickel.
- Because of its light weight, it is used in many weight-saving applications, particularly in the aircraft industry.
- Used in the manufacture and use of fireworks for signals and railroad flares.
- Magnesium casting has a good surface finish and is pressure tight making them ideal for this use. Uses are for engine housings, gearboxes, blowers, differential housings, hose pieces, landing wheels, portable tools, and certain parts of aircraft fuselage.

Magnesium alloy materials are used in sewing machines, typewriters, and textiles.

Magnesium Identification Tests

- Appearance Tests: Magnesium resembles aluminum in appearance. The polished surface is silver- white, but quickly oxidizes to a grayish film. Like aluminum, it is highly corrosion resistant and has a good strength-to-weight ratio, but is lighter in weight than aluminum. It has a very low kindling point and is not very weldable, except when it is alloyed with manganese and aluminum. Magnesium is distinguished from aluminum by the use of a silver nitrate solution. The solution does not react with aluminum but leaves a black deposit of silver on magnesium. Magnesium is produced in large quantities from sea water. It has excellent machinability, but special care must be used when machining because of its low kindling point.
- Fracture Test: Magnesium has a rough surface with a fine grain structure.
- Spark Test: No sparks are given off.
- Torch or Flame Test (Caution): Magnesium oxidizes rapidly when heated in open air, producing an oxide film which is insoluble in the liquid metal. A fire may result when magnesium is heated in the open atmosphere. As a safety precaution, magnesium should be melted in an atmosphere of inert gas.

Caution: Magnesium may ignite and burn when heated in the open atmosphere.

Manganese (Mn)

Pure manganese has a relatively high tensile strength but is very brittle. Manganese is used as an alloying agent in steel to deoxidize and desulfurize the metal. In metals other than steel, percentages of 1 to 15 percent manganese will increase the toughness and the hardenability of the metal involved.

Capabilities

Manganese can be welded, machined, and cold-worked.

Limitations

Austenitic manganese steels are best machined with cemented carbide, cobalt, and high-speed steel cutters.

Properties

- Tensile strength of 72,000 psi (496,440 kPa) (quenched).
- Brinell hardness number of 330.
- Specific gravity of 7.43.
- A melting point of 2270 °F (1243 °C); and is brittle. Manganese alloy has a tensile strength of 110,000 psi (758,450 kPa). Generally, manganese is highly polishable.
- Brittle.

Manganese alloy has a tensile strength of 110,000 psi (758,450 kPa).

Uses

Manganese is used mainly as an alloying agent in making steel to increase tensile strength. It is also added during the steel-making process to remove sulfur as a slag. Austenitic manganese steels are used for railroad track work, power shovel buckets, and rock crushers. Medium-carbon manganese steels are used to make car gears and axles.

Molybdenum (Mo)

Pure molybdenum has a high tensile strength and is very resistant to heat. It is principally used as an alloying agent in steel to increase strength, hardenability, and resistance to heat.

Capabilities

Molybdenum can be swaged, rolled, drawn, or machined.

Limitations

Molybdenum can only be welded by atomic hydrogen arc, or butt welded by resistance heating in vacuum. It is attacked by nitric acid, hot sulfuric acid, and hot hydrochloric acid.

Properties

- Tensile strength of 100,000 psi (689,500 kPa) (sheet) and 30,000 Psi (206,850 kPa) (wire).

- Brinell hardness number of 160 to 185.
- Specific gravity of 10.2.
- Melting point of 4800 °F (2649 °C).
- Retains hardness and strength at high temperatures.
- Corrosion resistant.

Uses

Molybdenum is used mainly as an alloy. Heating elements, switches, contacts, thermocouples, welding electrodes, and cathode ray tubes are made of molybdenum.

Monel

Monel metal is a nickel alloy of silver-white color containing about 67.00 percent nickel, 29.00 to 80.00 percent copper, 1.40 percent iron, 1.00 percent manganese, 0.10 percent silicon, and 0.15 percent carbon. In appearance, it resembles untarnished nickel. After use, or after contact with chemical solutions, the silver-white color takes on a yellow tinge and some of the luster. It has a very high resistance to corrosion and can be welded.

Monel is harder and stronger than either nickel or copper; acceptable substitute for steel in systems where corrosion resistance is the primary concern.

K-Monel vs. Monel

K-Monel was developed for greater strength and hardness than Monel; comparable to heat-treated steel; used for instrument parts that must resist corrosion.

Nickel (Ni)

Nickel is a hard, malleable, ductile tough metal. As an alloy, it will increase ductility, has no effect on grain size, lowers the critical point for heat treatment, aids fatigue strength, and increases impact values in low-temperature operations. Both nickel and nickel alloys are machinable and are readily welded by gas and arc methods. 85% of nickel is obtained from sulfide ores.

Excellent Corrosion Resistance is an Attractive Property of Nickel.

Color

The nickel color is silver.

Capabilities

Nickel alloys are readily welded by either the gas or arc methods. Nickel alloys can be machined, forged, cast, and easily formed.

Limitations

Nickel oxidizes very slowly in the presence of moisture or corrosive gasses.

Properties

- Strength: Pure nickel has tensile strength of 46,000 psi (317,170 kPa).
- Hardness: Brinell hardness number 220.
- Gravity: Specific gravity of 8.9.
- Melting point of 2650 °F (1454 °C).

Alloy Properties

- Strength: Tensile strength of 100,000 psi (689,500 kPa), and high strength and toughness at high temperatures.

Nickel Alloys

Nickel vs. Copper

Nickel has more strength but less ductility than copper.

Uses

Nickel is used in making alloys of both ferrous and nonferrous metal.

- Food processing and chemical equipment due to resistance to both alkalis and acids.
- Electrical resistance heating elements.
- Ornamental trim.
- Parts that must withstand elevated temperatures (produced from nickel-containing metal).
- Chromium alloys are used to make stainless steel.
- Plated on steel to provide a layer or surface for corrosion resistance.
- Steel alloy with high proportions of Nickel in Monel and Conel steel.

Nickel Identification Tests

- Appearance: Pure nickel has a greyish white.
- Fracture: The fracture surface of nickel is smooth and fine grained.
- Spark Test: In a spark test, nickel produces a very small amount of short, orange streaks which are generally wavy.

Tin (Sn)

Tin is used as an important alloy adding resistance to corrosion. Tin is a very soft, malleable, somewhat ductile, corrosion resistant metal having low tensile strength and high crystalline structure. It is used in coating metals to prevent corrosion.

Color

Tin is brilliant white in color with a yellow tinge.

Capabilities

Tin can be dying cast, cold worked (extruded), machined, and soldered.

Limitations

Tin is not weldable.

Properties

- Strength: Pure tin has tensile strength of 2800 psi (19,306 kPa).
- Gravity: Tin has a specific gravity of 7.29.
- Melting Point: Tin has a melting point of 450 °F (232 °C).
- Corrosion resistant.
- Malleable.
- Ductile.
- Can be rolled in thin sheets.
- Does not corrode in wet or dry conditions.

Babbitt alloy tin has the tensile strength of 10,000 psi (68,950 kPa) and Brinell hardness number of 30.

Uses

- Coating for steel.

- Food preservation as tin foil.
- As an alloying element:
  - With copper to produce tin brass and bronze.
  - With lead to produce solder.
  - With antimony and lead to form babbitt metal (Babbitt metal or bearing metal, is any of several alloys used for the bearing surface in a plain bearing).
- Tinning of brass and copper wire before conversion into cables and brass utensils.
- Useful protective coating for steel and iron and steel (due to corrosion resistant properties).
- Alloys for solder.
- Moisture proof packing (thin sheets).

Tin Identification Tests

- Appearance: Tin is silvery white in color.
- Fracture Test: The fracture surface of tin is silvery white and fairly smooth.
- Spark Test: Tin gives off no sparks in a spark test.
- Torch Test: Tin melts at 450 °F (232 °C), and will boil under the torch.

Titanium (Ti)

Titanium and its alloys offer excellent corrosion resistance to acids, chlorides, and salt. Titanium falls into a family of metals called reactive metals, which means that they have a high affinity for oxygen.

Titanium is a very soft, silvery-white, medium-strength metal having very good corrosion resistance. It has a high strength to weight ratio, and its tensile strength increases as the temperature decreases. Titanium has low impact and creep strengths, as well as seizing tendencies, at temperatures above 800 °F (427 °C).

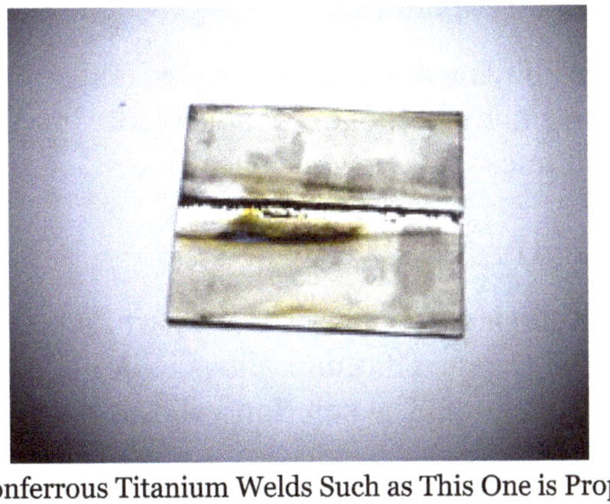

The Key To Nonferrous Titanium Welds Such as This One is Proper Shielding. Titanium is Stronger Than Steel But Light as Aluminum.

Capabilities

Titanium can be machined at low speeds and fast feeds; formal; spot-and seam- welded, and fusion welded using an inert gas.

Limitations

Titanium has low impact strength, and low creep strength at high temperatures (above 800 °F (427 °C)). It can only be cast into simple shapes, and it cannot be welded by any gas welding process because of its high attraction for oxygen. Oxidation causes this metal to become quite brittle. The inert gas welding process is recommended to reduce contamination of the weld metal.

Properties

- Tensile strength of 100,000 psi.
- Brinell hardness number of 200.
- Specific gravity of 4.5.
- Melting point of 3300 °F (1851 °C).
- Good corrosion resistance.

Alloy Titanium

- Brinell hardness number of 340.
- Tensile strength of 150,000 psi.
- High strength/weight ratio (twice that of aluminum alloy at 400 °F (204 °C)).

Uses

Titanium is a metal of the tin group which occurs naturally as titanium oxide or in other oxide forms. The free element is separated by heating the oxide with aluminum or by the electrolysis of the solution in calcium chloride. Its most important compound is titanium dioxide, which is used widely in welding electrode coatings. It is used as a stabilizer in stainless steel so that carbon will not be separated during the welding operation.

It is also used as an additive in alloying aluminum, copper, magnesium, steel, and nickel; making powder for fireworks; and in the manufacture of turbine blades, aircraft firewalls, engine nacelles, frame assemblies, ammunition tracks, and mortar base plates.

Titanium Metal Identification

- Appearance Test: Titanium is a soft, shiny, silvery-white metal burns in air and is the only element that burns in nitrogen. Titanium alloys look like steel and can be distinguished from steel by a copper sulphate. The solution will not react with titanium but will leave a coating of copper on steel.
- Spark Test: The sparks given off are large, brilliant white, and of medium length.

Tungsten (W)

Tungsten is a hard, heavy, nonmagnetic metal which will melt at approximately 6150°F (3400°C).

Capabilities

Tungsten can be hot drawn and cold.

Limitations

Tungsten is hard to machine, requires high temperatures for melting, and is produced by powdered metallurgy (sintering process). Singering is the process of compacting and forming a solid mass of material by heat or pressure.

Properties

- Melting point of 6170 ± 35 °F (3410 ± 19 °C)
- Ductility
- Tensile strength of 105,000 psi (723,975 kPa)
- Specific gravity of 19.32
- Thermal conductivity of 0.397
- Brinell hardness number of 38
- Dull white color

Tungsten Uses

Tungsten is used in making:

- Light bulb filaments
- Phonograph needles
- As an alloying agent in the production of high-speed steel, and projectiles
- Alloying agent in:
  - Non-consumable welding electrodes
  - Armor plate
  - Die and tool steels
  - Hard metal carbide cutting tools

It is one of the first metallic nanoscale powders.

Tungsten Identification Tests

- Appearance: Tungsten has a steel gray color.
- Spark Test: In a spark test Tungsten produces a very small volume of short, straight, orange streaks.

Vanadium (V)

Occurs with carbonaceous matter, free sulfur and iron pyrite. The color is silvery white.

Uses and Properties

- Castings can be produced using non-ferrous alloys such as aluminum and copper;
- Used when manufacturing alloy steel;
- Can be drawn into wires or hammered into any shape when heated to the appropriate temperature;
- Melting Point: 1710 °C;
- Specific Gravity: 5.67.

Zinc (Zn)

Zinc is a medium-low strength metal. It has a very low melting point and is somewhat heavy. Zinc is easy to machine but has a coarse grain.

Welding and Soldering

To prevent crystal cleavage, heat Zinc to about 180 °F (82 °C). If the heat input is controlled strictly and appropriately cleaned, Zinc can be soldered or welded.

Color

Blush-white.

Capabilities

Zinc can be welded, machined, cast and cold worked (extruded).

Limitations

Do not use Zinc die castings in continuous contact with steam.

Properties of Zinc

- Strength: Zinc has a tensile strength of 12,000 psi (82,740 kPa) (cast) and 27,000 psi (186,165 kPa) (rolled).

- Gravity: Zinc has a specific gravity of 7.1.
- Melting Point: Melting point of 790 °F (421 °C).
- Corrosion resistant.
- Brittle at 220 °F (104 °C).
- Boinging Point: The zinc boiling point is 940 °C.

Zinc Ores

- Blende (ZnS)
- Calamine ($ZnCo_3$)

Other Forms

- Zinc oxide and zinc sulfide: Used in rubber and paint.
- Zinc dust: used to manufacture chemicals and explosives.

Zinc Uses

- Galvanizing metal: Zinc is used on iron or steel in the form of a protective coating called galvanizing or electroplating. It is the largest use of zinc and is done by dipping the part in molten zinc or by electroplating it. Examples of items made in this way are bolts, nails, galvanized pipe, sheet metal, tubing, and wire.

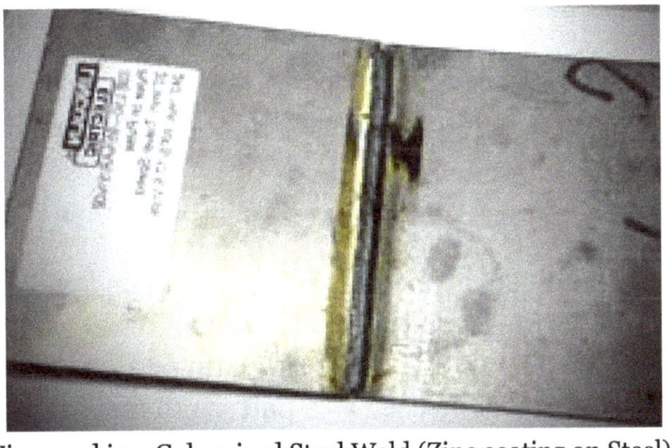

Zinc used in a Galvanized Steel Weld (Zinc coating on Steel).

- Alloying Element: Zinc is used as an alloying element in producing alloys such as brass and bronze.
- Parts due to Low Melting Point and High Fluidity: Typical parts made with zinc alloy die castings, toys, ornaments, building equipment, carburetor and fuel pump bodies, instrument panels, wet and dry batteries, fuse plugs, pipe organ pipes, munitions, cooking utensils, and flux.
- Rolled Sheets: Used for non-corrosive container lining and roof covering.

Zinc Identification Tests

- Appearance: Zinc and Zinc alloys are blue-white when polished. When they oxidized they are gray in color.
- Fracture Test: Fractures from Zinc appear to be granular.
- Spark Test: Zinc alloys and Zinc give off no sparks in a spark test.

Zinc Die Castings Identification Tests

- Appearance: Die castings are usually alloys of zinc, aluminum, magnesium, lead, and tin. They are light in weight, generally silvery white in color (like aluminum), and sometimes of intricate design. A die-cast surface is much smoother than that of a casting made in sand and is almost as smooth as a machined surface. Sometimes, die castings darkened by use may be mistaken for malleable iron when judged simply by looks, but the die casting is lighter in weight and softer.

This Spring Stop Contains the Nonferrous Metals Aluminum and Zinc.

- Fracture Test: The surface of a zinc die casting is white and has a slight granular structure.
- Spark Test: No sparks are given off by Zinc die castings.
- Torch Test: Zinc die-castings can be recognized by their low melting temperatures. The metal boils when heated with the oxyacetylene flame. A die casting, after a thorough cleaning, can be welded with a carburizing flame using tin or aluminum solders as filler metal. If necessary, the die-cast part can be used as a pattern to make a new brass casting.

White Metal Die-castings

These are usually made with alloys of aluminum, lead, magnesium, or tin. Except for those made of lead and tin, they are generally light in weight and are the color white.

White Metal Die Identification Tests

- Appearance Tests: The surface is smoother than casting made in sand.
- Fracture Test: Fractured surface is white and somewhat granular.

- Spark Test: In a spark test no sparks are given off.
- Torch Test: The metal boils under the torch due to low melting points.

Non-metals

A non-metal is simply an element that does not display the properties of a metal. It is not defined by what it is, but by what it is not. It doesn't look metallic, can't be made into a wire, pounded into shape or bent, doesn't conduct heat or electricity well, and doesn't have a high melting or boiling point.

The non-metals are in a minority on the periodic table, mostly located on the right-hand side of the periodic table. The exception is hydrogen, which behaves as a non-metal at room temperature and pressure and is found on the upper left corner of the periodic table. Under conditions of high pressure, hydrogen is predicted to behave as an alkali metal.

The non-metals are located on the upper right side of the periodic table. Non-metals are separated from metals by a line that cuts diagonally through the region of the periodic table containing elements with partially filled *p* orbitals. The halogens and noble gases are non-metals.

Properties of Non-metals

Physical Properties of Non-metals

- Physical State: Most of the non-metals exist in two of the three states of matter at room temperature: gases (oxygen) and solids (carbon). Only bromine exists as a liquid at room temperature.
- Non-Malleable and Ductile: Non-metals are very brittle, and cannot be rolled into wires or pounded into sheets.
- Conduction: They are poor conductors of heat and electricity.
- Luster: These have no metallic luster and do not reflect light.
- Melting and Boiling Points: The melting points of non-metals are *generally* lower than metals, but are highly variable.
- Seven non-metals exist under standard conditions as *diatomic molecules*: $H_2(g)$, $N_2(g)$, $O_2(g)$, $F_2(g)$, $Cl_2(g)$, $Br_2(l)$, $I_2(s)$.

Chemical Properties of Non-metals

Non-metals have a tendency to gain or share electrons with other atoms. They are electronegative in character. Non-metals, when reacting with metals, tend to gain electrons (typically attaining noble gas electron configuration) and become anions:

$$3Br_2(l) + 2Al(s) \rightarrow 2AlBr_3(s)$$

Compounds composed entirely of non-metals are covalent substances. They generally form acidic or neutral oxides with oxygen that that dissolve in water to form acids:

$$CO_2(g) + H_2O(l) \rightarrow \underset{\text{carbonic acid}}{H_2CO_3(aq)}$$

As you may know, carbonated water is slightly acidic (carbonic acid).

Non-metal oxides can combine with bases to form salts.

$$CO_2(g)+2NaOH(aq) \rightarrow Na_2CO_3(aq) + H_2O(l)$$

Metalloids

Metalloids are the smallest class of elements (the other two classes of elements are metals and non-metals). There are just six metalloids. In addition to silicon, they include boron, germanium, arsenic, antimony, and tellurium. Metalloids fall between metals and non-metals in the periodic table. They also fall between metals and non-metals in terms of their properties.

Chemical Properties of Metalloids

How metalloids behave in chemical interactions with other elements depends mainly on the number of electrons in their outer energy level of their atoms. Metalloids have from three to six electrons in their outer energy level. Boron, pictured in the figure below, is the only metalloids with just three electrons in its outer energy level. It tends to act like metals by giving up its electrons in chemical reactions. Metalloids with more than four electrons in their outer energy level (arsenic, antimony, and tellurium) tend to act like non-metals by gaining electrons in chemical reactions. Those with exactly four electrons in their outer energy level (silicon and germanium) may act like either metals or non-metals, depending on the other elements in the reaction.

Elemental boron

Physical Properties of Metalloids

Most metalloids have some physical properties of metals and some physical properties of non-metals. For example, metals are good conductors of both heat and electricity, whereas non-metals generally cannot conduct heat or electricity. And metalloids? They fall between metals and non-metals in their ability to conduct heat, and if they can conduct electricity, they usually can do so only at higher temperatures. Metalloids that can conduct electricity at higher temperatures are called semiconductors. Silicon is an example of a semiconductor. It is used to make the tiny electric circuits in computer chips. You can see a sample of silicon and a silicon chip in the figure.

Elemental silicon and a computer chip that is made with silicon.

Metalloids tend to be shiny like metals but brittle like non-metals. Because they are brittle, they may chip like glass or crumble to a powder if struck. Other physical properties of metalloids are more variable, including their boiling and melting points, although all metalloids exist as solids at room temperature.

Alloys

An alloy is the combination of metal with other chemical elements (metallic or non-metallic), forming a solution or chemical compound that retains metallic properties. Generally, a number of properties of the alloy are significantly different from those of its components. An alloy with two components is called a binary alloy; one with three is a ternary alloy; one with four is a quaternary alloy.

Steel is an extremely useful alloy, with iron as its major component and a carbon content between 0.02 and 1.7 percent by weight.

Among commonly known alloys is brass, bronze, and steel. Some alloys are named after their primary constituent. For example, the "silver" used in jewelry and the "aluminum" used as a structural building material are actually alloys. Alloys of gold are rated on a scale of carats—for instance, 14-carat gold is 58 percent gold. The term alloy is sometimes used loosely as a synonym for aluminum alloy, such as when referring to alloy wheels fitted on automobiles.

Some Common Alloys

Amalgam

Any alloy of mercury is called an amalgam. Most metals are soluble in mercury, but some (such as iron) are not. Amalgams are commonly used in dental fillings because they have been relatively cheap, easy to use, and durable. In addition, until recently, they have been regarded as safe. They are made by mixing mercury with silver, copper, tin, and other metals. The mercury content of dental fillings has recently stirred controversy, based on the potentially harmful effects of mercury.

Mercury amalgams have also been used in the process of mining gold and silver, because of the ease with which mercury amalgamates with them. In addition, thallium amalgam is used as the liquid material in thermometers, because it freezes at -58 °C, whereas pure mercury freezes at -38 °C.

Brass

Brass is the term used for alloys of copper and zinc in a solid solution. It has a yellow color, somewhat similar to gold. It was produced in prehistoric times, long before zinc was discovered, by melting copper with calamine, a zinc ore.

A decorative brass paperweight (left), along with zinc and copper samples.

The amount of zinc in brass varies from 5 to 45 percent, creating a range of brasses, each with unique properties. By comparison, bronze is principally an alloy of copper and tin. Despite this distinction, some types of brasses are called bronzes.

Brass is relatively resistant to tarnishing and is often used for decorative purposes. Its malleability and acoustic properties have made it the metal of choice for musical instruments such as the trombone, tuba, trumpet, and euphonium. Although saxophones and harmonicas are made out of brass, the saxophone is a woodwind instrument, and the harmonica, a free reed aerophone. In organ pipes designed as "reed" pipes, brass strips are used as the "reed."

Aluminum makes brass stronger and more corrosion-resistant. It forms a transparent, self-healing, protective layer of aluminum oxide (Al_2O_3) on the surface. Tin has a similar effect and finds its use especially in seawater applications (naval brasses). Combinations of iron, aluminum, silicon, and manganese make brass resistant to wear and tear.

Bronze

Bronze refers to a broad range of copper alloys, usually with tin as the main additive, but sometimes with other elements such as phosphorus, manganese, aluminum, or silicon. Typically, bronze is about 60 percent copper and 40 percent tin.

Assorted ancient bronze castings found as part of a cache.

The use of bronze was particularly significant for early civilizations, leading to the name "Bronze Age." Tools, weapons, armor, and building materials such as decorative tiles were made of bronze, as they were found to be harder and more durable than their stone and copper predecessors. In early use, the natural impurity arsenic sometimes created a superior natural alloy, called "arsenical bronze."

Though not as strong as steel, bronze is superior to iron in nearly every application. Bronze develops a patina (a green coating on the exposed surface), but it does not oxidize beyond the surface. It is considerably less brittle than iron and has a lower casting temperature. Several bronze alloys resist corrosion (especially by seawater) and metal fatigue better than steel; they also conduct heat and electricity better than most steels.

Bronze has myriad uses in industry. It is widely used today for springs, bearings, bushings, and similar fittings, and is particularly common in the bearings of small electric motors. It is also widely used for cast metal sculpture and is the most popular metal for top-quality bells and cymbals.

Commercial bronze, otherwise known as brass, is 90 percent copper and 10 percent zinc. It contains no tin.

Pewter

Pewter is traditionally composed of 85 to 99 percent tin, with the remainder consisting of copper, which acts as a hardener. Lead is added to lower grades of pewter, giving a bluish tint.

Pewter plate

Traditionally, there were three grades of pewter: fine, for eatingware, with 96 to 99 percent tin and 1 to 4 percent copper; trifle, also for eating and drinking utensils but duller in appearance, with 92 percent tin, 1 to 4 percent copper, and up to 4 percent lead; and lay or ley metal, not for eating or drinking utensils, which could contain up to 15 percent lead. Modern pewter mixes the tin with copper, antimony, and bismuth rather than lead.

Physically, pewter is a bright, shiny metal that is similar in appearance to silver. Like silver, it oxidizes to a dull gray over time if left untreated. It is a very malleable alloy, being soft enough to carve with hand tools. It also takes good impressions from punches or presses. Given this inherent softness and malleability, pewter cannot be used to make tools. Some types of pewter pieces, such as candlesticks, were turned on a metal lathe, and these items are sometimes referred to as "holloware." Pewter has a low melting point (around 225 to 240 °C), depending on the exact mixture of metals. Duplication by casting gives excellent results.

The use of pewter was common from the Middle Ages up until the various developments in glass-making during the eighteenth and nineteenth centuries. Pewter was the chief tableware until the making of china. With the mass production of glass products, glass has universally replaced pewter in daily life. Today, pewter is mainly used for decorative objects such as collectible statuettes and figurines, replica coins, and pendants.

Nickel Silver (German Silver)

Nickel silver is an alloy of copper, nickel, and often (but not always) zinc. It is named for its silvery appearance and contains no elemental silver. Other common names for this alloy are German silver, paktong, new silver, and alpacca (or alpaca).

Many different formulations of alloys fall in the general category of "nickel silver." Besides containing copper, nickel, and zinc, some formulations may include antimony, tin, lead, or cadmium. A representative industrial formulation (Alloy No.752) is 65 percent copper, 18 percent nickel, and 17 percent zinc. In metallurgical science, such alloys would be more properly termed nickel bronze. Some nickel silver alloys, especially those containing high proportions of zinc, are stainless.

The earliest use of nickel silver appears to have been in China. It became known to the West from imported wares called Paktong or Pakfong, where the silvery metal color was used to imitate sterling silver. It was discovered to be an alloy composed of copper, nickel, and zinc in the eighteenth century.

Nickel silver first became popular as a base metal for silver-plated cutlery and other silverware, notably the electroplated wares called "E.P.N.S." (electro-plated nickel silver). It is used in zippers, costume jewelry, and musical instruments (such as cymbals). After about 1920, its use became widespread for pocketknife bolsters, due to its machinability and corrosion resistance. In some countries, it is used in the production of coins. Its industrial and technical uses include marine fittings and plumbing fixtures for its corrosion resistance, and heating coils for its high electrical resistance.

Steel

Steel is an alloy composed mainly of iron, with carbon content between 0.02 and 1.7 percent by weight. Carbon is the most cost-effective alloying material for iron, but many other alloying elements are also used. Carbon and other elements act as hardening agents, preventing iron atoms in the crystal lattice from sliding past one another.

By varying the amount of alloying elements and their distribution in the steel, one can control its qualities such as hardness, elasticity, ductility, and tensile strength. Steel with increased carbon content can be made harder and stronger than iron, but it is also more brittle. The maximum solubility of carbon in iron is 1.7 percent by weight, occurring at 1130 °C. Higher concentrations of carbon or lower temperatures produce cementite, which reduces the material's strength. Alloys with higher carbon content than this are known as cast iron because of their lower melting point. Steel should also be distinguished from wrought iron, with little or no carbon (usually less than 0.035 percent).

Currently, there are several classes of steels in which carbon is replaced with other alloying materials, and carbon, if present, is undesired. More recently, steels have been defined as iron-based alloys that can be plastically formed—pounded, rolled, and so forth.

List of Alloys

This is a list of alloys grouped by the main metallic constituent, in order of increasing atomic number of the main metal. Under these headings, the alloys are in no particular order. Some of the main alloying elements are optionally listed after the alloy names.

Alloys of Aluminum

Al-Li (lithium, mercury), Duralumin (copper), Nambe (seven undisclosed metals), Magnox (magnesium oxide), Zamak (zinc, magnesium, copper), Silumin (silicon).

Alloys of Potassium

NaK (sodium)

Alloys of Iron

- Steel (carbon)
    - Stainless steel (chromium, nickel): AL-6XN, Alloy 20, Celestrium, Marine grade stainless, Martensitic stainless steel, surgical stainless steel (chromium, molybdenum, nickel).

- Silicon steel (silicon), Tool steel (tungsten or manganese), Bulat steel Chromoly (chromium, molybdenum), Crucible steel Damascus steel, HSLA steel, High speed steel, Maraging steel, Reynolds 531, Wootz steel.

- Iron: Anthracite iron (carbon), Cast iron (carbon), Pig iron (carbon), Wrought iron (carbon).
- Fernico (nickel, cobalt), Elinvar (nickel, chromium), Invar (nickel), Kovar (cobalt), Spiegeleisen (manganese, carbon, silicon).
- Ferroalloys: Ferroboron, Ferrochrome, Ferromagnesium, Ferromanganese, Ferromolybdenum, Ferronickel, Ferrophosphorus, Ferrotitanium, Ferrovanadium, Ferrosilicon.

Alloys of Cobalt

Megallium, Stellite (chromium, tungsten, carbon), Talonite, Alnico, Vitallium.

Alloys of Nickel

Nickel silver/German silver (copper, zinc), Chromel (chromium), Hastelloy (molybdenum, chromium, sometimes tungsten), Inconel (chromium, iron), Mu-metal (iron), Monel metal (copper, nickel, iron, manganese), Nichrome (chromium, iron, nickel), Nicrosil (chromium, silicon, magnesium), Nisil (silicon), Nitinol (titanium, shape memory alloy), Copper-Nickel (bronze, copper).

Alloys of Copper

- Beryllium copper (beryllium), Billon (silver).
- Brass (zinc): Calamine brass (zinc), Chinese silver (zinc), Gilding metal (zinc), Muntz metal (zinc), Pinchbeck (zinc), Prince's metal (zinc), Tombac (zinc).
- Bronze (tin, aluminum, or any other element): Aluminum bronze (aluminum), Bell metal (tin), Guanín, Gunmetal (tin, zinc), Phosphor bronze (tin and phosphorus), Ormolu (Gilt Bronze) (zinc), Speculum metal (tin).
- Constantan (nickel),Corinthian brass (gold, silver), Cunife (nickel, iron), Cupronickel (nickel), Cymbal alloys (Bell metal) (tin), Devarda's alloy (aluminum, zinc), Hepatizon (gold, silver), Heusler alloy (manganese, tin), Manganin (manganese, nickel), Nickel silver (nickel), Nordic gold (aluminum, zinc, tin), Shakudo (gold), Tumbaga (gold).

Alloys of Gallium: Galinstan.

Alloys of Silver: Sterling silver (copper), Britannia silver (copper).

Alloys of Tin: Britannium (copper, antimony), Pewter (lead, copper), Solder (lead, antimony).

Rare Earth Alloys: Mischmetal (various rare earths).

Alloys of Gold: Corinthian brass (copper), Electrum (silver, copper), Tumbaga (copper), Rose gold (copper), White gold.

Alloys of Mercury: Amalgam.

Alloys of Lead: Solder (tin), Terne (tin), Type metal (tin, antimony).

Alloys of Bismuth: Wood's metal, Rose metal, Field's metal, Cerrobend.

Alloys of Zirconium: Zircaloy.

Ceramics

Ceramics are an incredibly diverse family of materials whose member's span traditional ceramics (such as pottery and refractories) to the modern day engineering ceramics (such as alumina and silicon nitride) found in electronic devices, aerospace components and cutting tools.

Ceramics exhibit very strong ionic and covalent bonding (stronger than the metallic bond) and this confers the properties commonly associated with ceramics: high hardness, high compressive strength, low thermal and electrical conductivity and chemical inertness.

This strong bonding also accounts for the less attractive properties of ceramics, such as low ductility and low tensile strength. The wider range of properties, however, is not widely appreciated. For example, whilst ceramics are perceived as electrical and thermal insulators, ceramic oxides (initially based on Y-Ba-Cu-O) are the bases for high temperature superconductivity. Diamond, beryllia and silicon carbide have a higher thermal conductivity than aluminium or copper.

Control of the microstructure can overcome inherent stiffness to allow the production of ceramic springs, and ceramic composites have been produced with a fracture toughness about half that of steel.

Composition and Classification

Traditional ceramics have been mainly silicate-based. Advanced ceramics are made from various other types of materials as well. Depending on their composition, they are classified as oxides, non-oxides, and composites.

- Oxides: alumina, zirconia.
- Non-oxides: Carbides, borides, nitrides, silicides, and silicates.
- Composites: Particulate reinforced combinations of oxides and non-oxides.

The materials in each of these classes can have unique properties.

Ceramic Materials and their uses

- Barium titanate (often mixed with strontium titanate), which has properties called *ferroelectricity* and *piezoelectricity*, is widely used in electromechanical devices known as *transducers*, as well as in ceramic capacitors and data storage elements.

A small sample of the high-temperature superconductor BSCCO-2223 (bismuth strontium calcium copper oxide); the two lines in the background are 1 mm apart.

- Bismuth strontium calcium copper oxide (BSCCO) is a high-temperature superconductor.
- Boron carbide (B_4C) is used in some types of personal, helicopter, and tank armor.
- Boron nitride takes on physical forms that are similar to those of carbon: a graphite-like form used as a lubricant, and a diamond-like one used as an abrasive.
- Bricks, which are mostly aluminum silicates, are used for construction.
- "Cermet," the name given to a composite of ceramic and metal, is used to produce capacitors, resistors, and other electronic parts for high-temperature applications.
- Earthenware is often made from clay, quartz, and feldspar.
- Ferrite (Fe_3O_4), which is ferrimagnetic, is used in the core of electrical transformers and in magnetic core memory.
- Lead zirconate titanate (also known as PZT) is a ferroelectric and piezoelectric material and has similar uses as barium titanate.
- Magnesium diboride (MgB_2) is an unconventional superconductor.
- Porcelain, which usually contains the clay mineral kaolinite, is used to make decorative and household items.
- Silicon carbide (SiC) is used as an abrasive, a refractory material, and a "susceptor" that helps cook food in microwave furnaces.
- Silicon nitride (Si_3N_4) is used as an abrasive powder.
- Steatite (a type of soapstone) is used as an electrical insulator.
- Uranium oxide (UO_2) is used as fuel in nuclear reactors.
- Yttrium barium copper oxide ($YBa_2Cu_3O_{7-x}$) is another high-temperature superconductor.

- Zinc oxide (ZnO) is a semiconductor and is used in the construction of varistors (a class of electrical resistors).
- Zirconia, which in the pure form undergoes many phase changes when heated, can be chemically "stabilized" in several different forms. Most ceramic knife blades are made of this material. Also, as it is a good conductor of oxygen ions, it could be useful in fuel cells.

Other Applications

In the early 1980s, Toyota researched production of a ceramic engine that could run at a temperature above 6,000 °F (3,300 °C). Ceramic engines do not require a cooling system and hence permit major weight reduction and higher fuel efficiency. In a conventional metallic engine, much of the energy released by combustion of the fuel must be dissipated as waste heat, to prevent the metallic parts from melting. Despite these desirable properties, such engines are not being mass-produced because it is difficult to manufacture ceramic parts with the requisite precision and durability. Imperfections in the ceramic material can lead to cracks and potentially dangerous equipment failure.

Efforts are being made to develop ceramic parts for gas turbine engines. Currently, even blades made of advanced metal alloys for the hot section of an engine require cooling and careful limiting of operating temperatures. Turbine engines made with ceramics could operate more efficiently, giving aircraft greater range and payload for a set amount of fuel.

Since the late 1990s, highly specialized ceramics, usually based on boron carbide, have been used in ballistic armored vests to repel large-caliber rifle fire. Such plates are commonly known as "small-arms protective inserts" (SAPI). Similar technology is used to armor the cockpits of some military airplanes because of the lightness of the material.

Recent advances in ceramics include bio-ceramics such as dental implants and synthetic bones. Hydroxyapatite, the natural mineral component of bone, has been synthesized from a number of biological and chemical sources and can be formed into ceramic materials. Orthopedic implants made from these materials bond readily to bone and other tissues in the body without rejection or inflammatory reactions.

Hydroxyapatite ceramics, however, are usually porous and lack mechanical strength. They are therefore used to coat metal orthopedic devices, to aid in forming a bond to bone, or as bone fillers. They are also used as fillers for orthopedic plastic screws to help reduce inflammation and increase absorption of the plastic materials. Work is being done to make strong, dense, nano-crystalline hydroxyapatite ceramics for orthopedic weight-bearing devices, replacing metal and plastic materials. Ultimately, these ceramic materials, with the incorporation of proteins called *collagens*, may be used to make synthetic bones.

Properties of Ceramics

Mechanical Properties

Ceramic materials can be crystalline or amorphous (lacking a definite structure). They are hard and resistant to high temperatures. In addition, they tend to be porous. The pores and other microscopic imperfections act as stress concentrators. As a result, ceramic materials are brittle—when placed under stress, they fracture rather than becoming deformed. The study of ceramics consists to a large extent of ways to accentuate the strengths and mitigate the limitations of the materials, as well as to develop new uses for these materials.

Electrical Properties

Semiconductivity

A number of ceramics are semiconductors. Most of these are oxides of transition metals, such as zinc oxide.

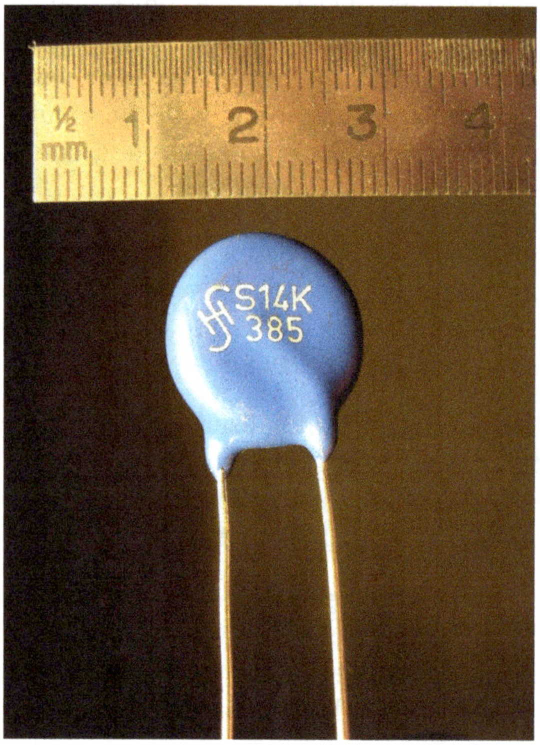

A 385-volt metal oxide varistor.

One common use of these semiconductors is for varistors. These are electrical resistors with the unusual property of "negative resistance." Once the voltage across the device reaches a certain threshold, a change in the electrical structure of the material causes its electrical resistance to drop from several megaohms down to a few hundred ohms. As a result, these materials can dissipate a lot of energy. In addition, they self-reset—after the voltage across the device drops below a threshold, its resistance returns to being high.

This property makes them ideal for surge-protection applications. The best demonstration of their ability can be found in electrical substations, where they are employed to protect the infrastructure from lightning strikes. They have rapid response, require low maintenance, and do not appreciably degrade from use.

When various gases are passed over a polycrystalline ceramic, its electrical resistance changes. Based on this property, semiconducting ceramics are used to make inexpensive gas sensors.

Superconductivity

Under some conditions, such as extremely low temperatures, some ceramics exhibit superconductivity. The exact reason for this property is not known, but there are two major families of superconducting ceramics.

Piezoelectricity, Pyroelectricity and Ferroelectricity

Many ceramic materials exhibit the property of *piezoelectricity*. A piezoelectric material develops a voltage difference between two faces when compressed or made to vibrate. This property links electrical and mechanical responses. Such ceramics are used in digital watches and other electronics that rely on quartz resonators. In these devices, electricity is used to produce a mechanical motion (powering the device) and the mechanical motion is in turn used to generate an electrical signal.

The piezoelectric effect is generally stronger in materials that also exhibit *pyroelectricity*. Such materials generate an electrical potential when heated or cooled. All pyroelectric materials are also piezoelectric. Thus, pyroelectric materials can be used to interconvert between thermal, mechanical, and electrical forms of energy. Such materials are used in motion sensors, where the tiny rise in temperature from a warm body entering a room is enough to produce a measurable voltage in the crystal.

Pyroelectricity, in turn, is observed most strongly in materials that also display the *ferroelectric effect*. Ferroelectric materials have a spontaneous polarization (formation of an electric dipole) that can be reversed by applying an electric field. Pyroelectricity is a necessary consequence of ferroelectricity.

Barium titanate and lead zirconate titanate have both piezoelectric and ferroelectric properties. They are used in the manufacture of capacitors, high-frequency loudspeakers, transducers for sonar, and actuators for atomic force and scanning tunneling microscopes.

Change of Electrical Properties with Temperature

Some crystalline semiconducting ceramics (mostly mixtures of heavy metal titanates) can conduct electricity as long as they are below a certain "transition" temperature. When heated above that temperature, their "grain boundaries" (boundaries between the little crystals that make up the material) suddenly become insulating, breaking the circuit. Such ceramics are used as self-controlled heating elements in, for example, the rear-window defrosts circuits of most automobiles.

Processing of Ceramic Materials

Non-crystalline ceramics, being glasses, are usually formed from melts. The glass is shaped when fully molten or when in a state of toffee-like viscosity, by methods such as casting or blowing to a mold. If the material becomes partly crystalline by later heat treatments, the resulting material is known as a "glass ceramic."

Crystalline ceramic materials are generally processed by one of two approaches: (a) the ceramic is made in the desired shape by reaction in situ, or (b) powders are "formed" into the desired shape and then heated ("sintered") until the particles adhere to one another and produce a solid body. Some methods use a hybrid of the two approaches.

In Situ Manufacturing

This method is most commonly used for producing cement and concrete. In this case, the dehydrated powders are mixed with water, which starts what are called *hydration reactions*. As a result, long, interlocking crystals begin to form around the aggregates. Over time, a solid ceramic is produced.

The biggest problem with this method is that most reactions are so fast that good mixing is not possible, preventing large-scale construction. On the other hand, small-scale systems can be made by "deposition" techniques—various materials (reactants) are introduced above a substrate, and made to react to form the ceramic on the substrate. This process borrows techniques from the semiconductor industry and is very useful for making ceramic coatings.

Sintering-based Approach

The term *sintering* refers to the process of heating a material at a temperature below its melting point, so that the particles adhere to one another. In a sintering-based approach, the object (called a "green body") that is prepared from a powder is initially held together loosely, but it is hardened by baking in a kiln. The pores in the object close up, so that the body shrinks and is converted to a denser, stronger product. There is virtually always some porosity left, but the real advantage of this method is that the green body can be produced in a wide variety of ways and then sintered.

Slurry can be used in place of a powder, then cast into a desired shape, dried, and sintered. This approach is used for traditional pottery, in which a plastic mixture is worked by hand.

When a mixture of different materials is used to form a ceramic, the sintering temperature is sometimes above the melting point of a minor component, leading to "liquid phase" sintering. In this case, the sintering time is shorter than for solid phase sintering.

Refractories

Refractories are ceramic materials designed to withstand the very high temperatures (in excess of

1,000 °F [538 °C]) encountered in modern manufacturing. More heat-resistant than metals, they are used to line the hot surfaces found inside many industrial processes.

In addition to being resistant to thermal stress and other physical phenomena induced by heat, refractories can withstand physical wear and corrosion caused by chemical agents. Thus, they are essential to the manufacture of petrochemical products and the refining of gasoline.

Composition and Processing

The composition and processing of ceramic refractories vary widely according to the application and the type of refractory. Most refractories can be classified on the basis of composition as either clay-based or nonclay-based. In addition, they can be classified as either acidic (containing silica [SiO_2] or zirconia [ZrO_2]) or basic (containing alumina [Al_{203}] or alkaline-earth oxides such as lime [CaO] or magnesia [MgO]). Among the clay-based refractories are fireclay, high-alumina, and mullite ceramics. There is a wide range of nonclay refractories, including basic, extra-high alumina, silica, silicon carbide, and zircon materials. Most clay-based products are processed in a manner similar to other traditional ceramics such as structural clay products; e.g., stiff-mud processes such as press forming or extrusion are employed to form the ware, which is subsequently dried and passed through long tunnel kilns for firing. Firing induces partial vitrification, or glass formation, which is a liquid-sintering process that binds particles together. Nonclay-based refractories, on the other hand, are bonded using techniques reserved for advanced ceramic materials. For instance, extra-high alumina and zircon ceramics are bonded by transient-liquid or solid-state sintering, basic bricks are bonded by chemical reactions between constituents, and silicon carbide is reaction-bonded from silica sand and coke.

Clay-based Refractories

The composition and properties of the clay-based refractories are described. Most are produced as preformed brick. Much of the remaining products are so-called monolithics, materials that can be formed and solidified on-site. This category includes mortars for cementing bricks and mixes for ramming or gunning (spraying from a pressure gun) into place. In addition, lightweight refractory insulation can be made in the form of fibreboards, blankets, and vacuum-cast shapes.

Fireclay

The workhorses of the clay-based refractories are the so-called fireclay materials. These are made from clays containing the aluminosilicate mineral kaolinite ($Al_2[Si_2O_5][OH]_4$) plus impurities such as alkalis and iron oxides. The alumina content ranges from 25 to 45 percent. Depending upon the impurity content and the alumina-to-silica ratio, fireclays are classified as low-duty, medium-duty, high-duty, and super-duty, with use temperature rising as alumina content increases. Fireclay bricks, or firebricks, exhibit relatively low expansion upon heating and are therefore moderately resistant against thermal shock. They are fairly inert in acidic environments but are quite reactive in basic environments. Fireclay bricks are used to line portions of the interiors of blast furnaces, blast-furnace stoves, and coke ovens.

High Alumina

High-alumina refractories are made from bauxite, a naturally occurring material containing aluminum hydroxide ($Al[OH]_3$) and kaolinitic clays. These raw materials are roasted to produce a mixture of synthetic alumina and mullite (an aluminosilicate mineral with the chemical formula $3Al_2O_3 \cdot 2SiO_2$). By definition high-alumina refractories contain between 50 and 87.5 percent alumina. They are much more robust than fireclay refractories at high temperatures and in basic environments. In addition, they exhibit better volume stability and abrasion resistance. High-alumina bricks are used in blast furnaces, blast-furnace stoves, and liquid-steel ladles.

Mullite

Mullite is an aluminosilicate compound with the specific formula $3Al_2O_3 \cdot 2SiO_3$ and an alumina content of approximately 70 percent. It has a melting point of 1,850° C (3,360 °F). Various clays are mixed with bauxite in order to achieve this composition. Mullite refractories are solidified by sintering in electric furnaces at high temperatures. They are the most stable of the aluminosilicate refractories and have excellent resistance to high-temperature loading. Mullite bricks are used in blast-furnace stoves and in the forehearth roofs of glass-melting furnaces.

Non-clay-based Refractories

Nonclay refractories such as those described below are produced almost exclusively as bricks and pressed shapes, though some magnesite-chrome and alumina materials are fuse-cast into molds. The usual starting materials for these products are carbonates or oxides of metals such as magnesium, aluminum, and zirconium.

Basic Refractories

Basic refractories include magnesia, dolomite, chrome, and combinations of these materials. Magnesia brick is made from periclase, the mineral form of magnesia (MgO). Periclase is produced from magnesite (a magnesium carbonate, $MgCO_3$), or it is produced from magnesium hydroxide ($Mg[OH]_2$), which in turn is derived from seawater or underground brine solutions. Magnesia bricks can be chemically bonded, pitch-bonded, burned, or burned and then pitch-impregnated.

Dolomite refractories take their name from the dolomite ore, a combination of calcium and magnesium carbonates ($CaCO_3 \cdot MgCO_3$), from which they are produced. After burning they must be impregnated with tar or pitch to prevent rehydration of lime (CaO). Chrome brick is made from chromium ores, which are complex solid solutions of the spinel type (a series of oxide minerals including chromite and magnetite) plus silicate gangue, or impurity phases.

All the basic refractories exhibit outstanding resistance to iron oxides and the basic slags associated with steelmaking—especially when they incorporate carbon additions either as flakes or as residual carbon from pitch-bonding or tar-impregnation. For this reason they find wide employment in the linings of basic oxygen furnaces, electric furnaces, and open-hearth furnaces. They also are used to line the insides of copper converters.

Extra-high Alumina

Extra-high alumina refractories are classified as having between 87.5 and 100 percent Al_2O_3 content. The alumina grains are fused or densely sintered together to obtain high density. Extra-high alumina refractories exhibit excellent volume stability to over 1,800 °C (3,275 °F).

Silica

Silica refractories are made from quartzites and silica gravel deposits with low alumina and alkali contents. They are chemically bonded with 3–3.5 percent lime. Silica refractories have good load resistance at high temperatures, are abrasion-resistant, and are particularly suited to containing acidic slags. Of the various grades—coke-oven quality, conventional, and super-duty—the super-duty, which has particularly low impurity contents, is used in the superstructures of glass-melting furnaces.

Zircon

Refractories made of zircon (a zirconium silicate, $ZrSiO_4$) also are used in glass tanks because of their good resistance to the corrosive action of molten glasses. They possess good volume stability for extended periods at elevated temperatures, and they also show good creep resistance (i.e., low deformation under hot loading).

Silicon Carbide

Silicon carbide (SiC) ceramics are made by a process referred to as reaction bonding, invented by the American Edward G. Acheson in 1891. In the Acheson process, pure silica sand and finely divided carbon (coke) are reacted in an electric furnace at temperatures in the range of 2,200°–2,480 °C (4,000°–4,500 °F). SiC ceramics have outstanding high-temperature load-bearing strength and dimensional stability. They also exhibit great thermal shock resistance because of their high thermal conductivity. (In this case, high thermal conductivity prevents the formation of extreme temperature differences between inner and outer layers of a material, which frequently are a source of thermal expansion stresses.) Therefore, SiC makes good kiln furniture for supporting other ceramics during their firing.

Other Non-clay-based Refractories

Other refractories produced in smaller quantities for special applications include graphite (a layered, multicrystalline form of carbon), zirconia (ZrO_2), forsterite (Mg_2SiO_4), and combinations such as magnesia-alumina, magnesite-chrome, chrome-alumina, and alumina-zirconia-silica. Alumina-zirconia-silica (AZS), which is melted and cast into molds or directly into the melting tanks of glass furnaces, is an excellent corrosion-resistant refractory that does not release impurities into the glass melt. AZS is also poured to make tank blocks (also called soldier blocks or sidewall blocks) used in the construction and repair of glass furnaces.

Glasses

Glasses are a unique range of ceramic materials defined principally by their atomic structure.

Glasses do not exhibit the ordered crystalline structure of most other ceramics but instead have a highly disordered amorphous structure. This gives them very different properties to other crystalline ceramics.

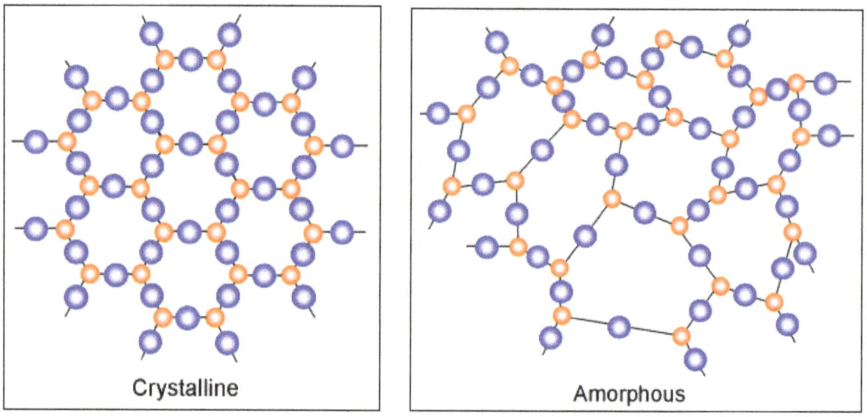

The most widely used glasses are silicate glasses, formed from silica, SiO_2. Silica consists of a 3D network of tetrahedra where every corner oxygen atom is shared with the adjacent tetrahedron. This SiO_2 tetrahedral unit is also incorporated into chains and sheets (clays), forming different ceramics.

Pure silica can be made to exist as a glass, and is called fused silica. A glass is a material that has hardened and become rigid without crystallising, making it amorphous. Silicate glasses are the most widely used glasses.

All of the most important glasses are based on silica SiO_2. Two glasses of special interest are soda-lime glass, used in windows and bottles, with composition $70SiO_2$. $10CaO$. $15Na_2O$, and borosilicate glass, used in cooking and chemical glassware, with composition $80SiO_2$. $15B_2O_3$. $5Na_2O$.

The ability of an oxide to form a glass upon cooling, i.e. an amorphous structure, depends on the structural relationship between the oxygen atoms and the cations of the oxide compound. In order to achieve a glass structure the oxide cations will bind with the oxygen atoms to form a tetrahedral network.

Oxides that form glasses are known as network formers. Common examples are SiO_2, B_2O_3, GeO_2, and P_2O_5.

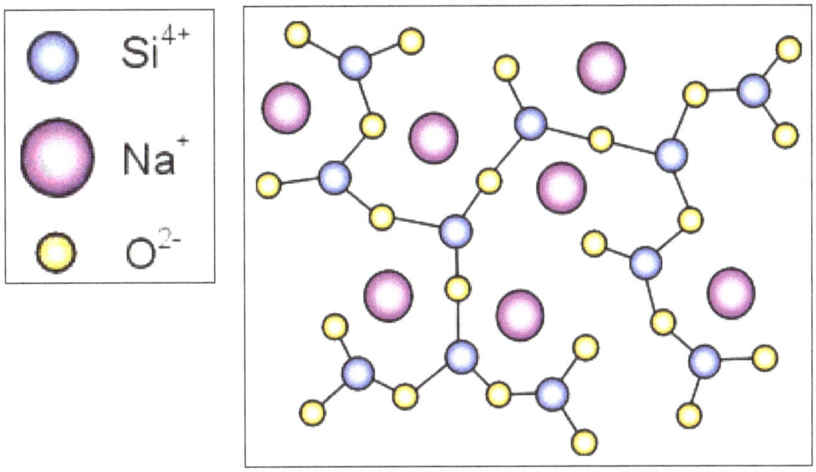

Common examples of intermediates are TiO_2, ZnO, PbO, and Al_2O_3.

Certain other oxides can be added to substitute for Si atoms in the tetrahedral structure. These oxides become part of the network and act as a stabiliser.

The oxides are known as intermediates and, in a similar manner to network modifiers, lower the melting point and viscosity of the glass thereby allowing it to be worked at lower temperatures.

Glass with added sodium as a network modifier.

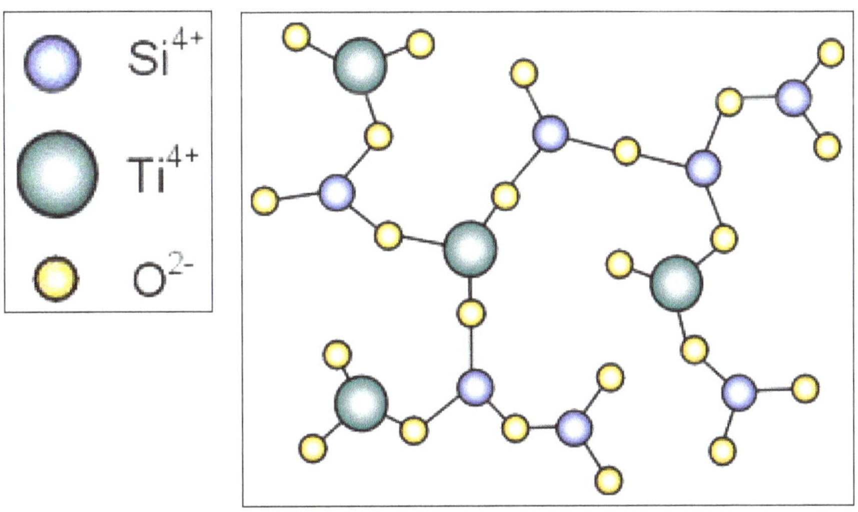

Glass with added titanium as an intermediate.

Properties

Due to its amorphous structure, a glass at room temperature is essentially a very viscous liquid. Viscous flow takes place in glass but so slow as to be on a geological time scale.

The viscosity of glass changes with temperature. There are four standard temperature points that are used in glass working to define the viscosity. These points are known as strain, annealing, softening, and working. The strain point is the maximum temperature at which a glass can be used for structural applications without undergoing creep.

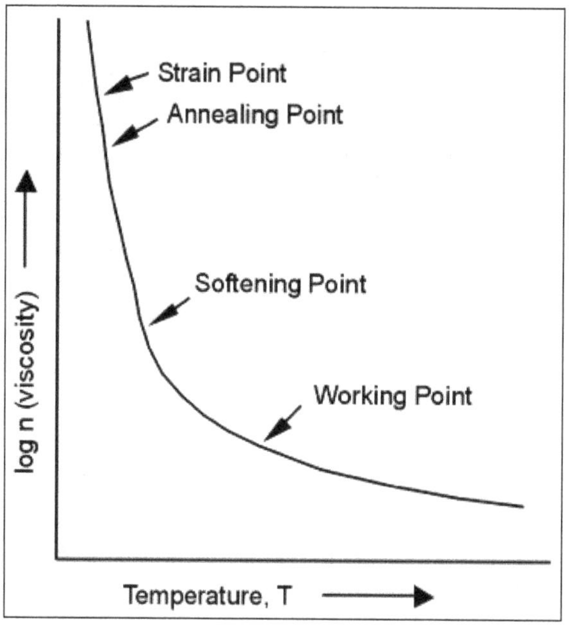

The majority of glass forming operations occurs between the softening point and the working point. At the softening point the glass will yield with a small amount of force while at the working point the glass has the viscosity of honey.

The annealing point is the temperature to which a glass may be heated after working to relieve any internal stresses that arose as a result of the forming process.

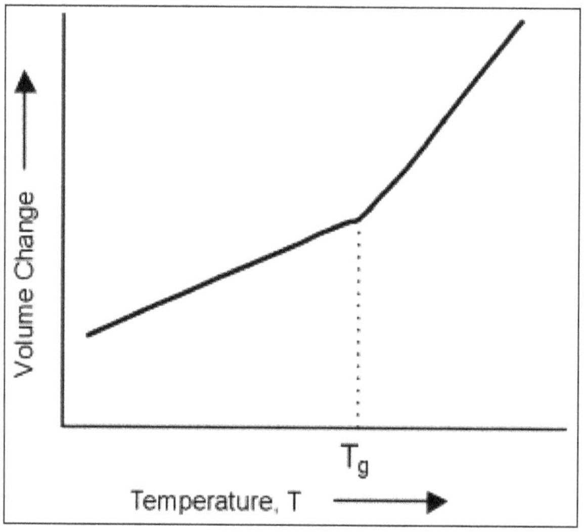

As well as an increase in viscosity, glass undergoes a thermal contraction when it is cooled down.

When cooling at high temperatures the glass undergoes a rapid decrease in volume. Glass at this stage has a soft rubbery texture. At some point the glass transforms to a hard brittle texture and the glass contracts at a slower rate.

The temperature at which this transformation occurs is known as the Glass Transition Temperature, T_g.

Forming

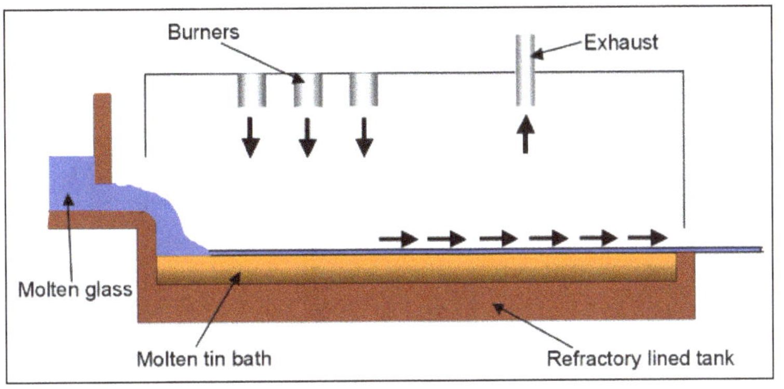

Almost all flat glass manufactured today is done so through a float bath process.

In this process molten glass is moved out of a furnace and floated on the surface of a bath of molten tin. The glass sheet slowly cools as it moves along the bath within a controlled atmosphere.

Float Bath Process

Once the sheet of glass is sufficiently hard it is moved off the bath and passed through an annealing furnace to remove residual stresses.

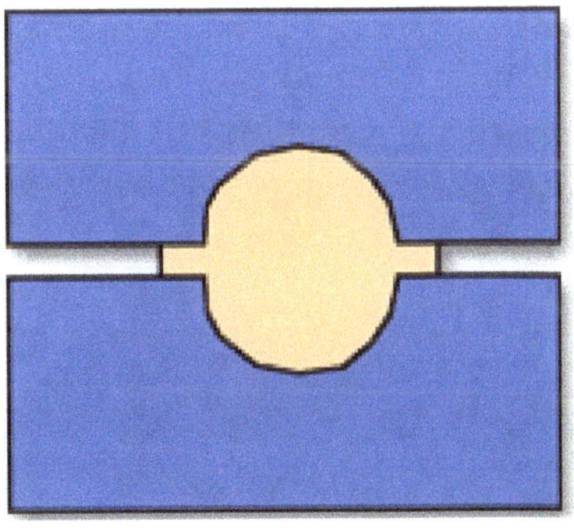

Thick walled pieces of glass may be hot pressed. In this process a molten slug of glass is placed between two mould surfaces and the glass is clamped into the desired shape.

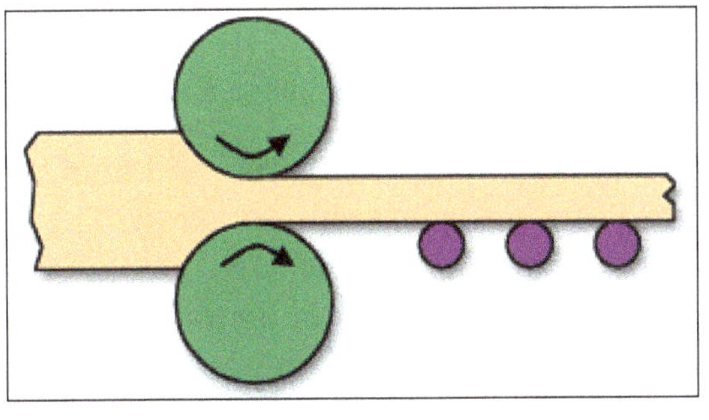

Sheet glass may also be formed through a rolling process whereby the molten glass is passed between two circulating rollers. This process does not produce the high quality of the molten bath process.

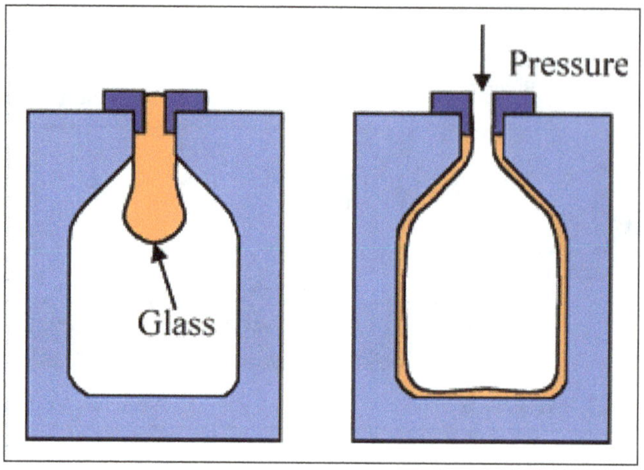

The formation of glass bottles and containers is usually achieved through a blow moulding process where the glass is placed in a mould and forced to conform to the mould walls under high air pressure.

Composites

Composite materials (or composites for short) are engineered materials made from two or more constituent materials with significantly different physical or chemical properties and which remain separate and distinct on a macroscopic level within the finished structure.

A cloth of woven carbon fiber filaments, a common element in composite materials.

The most primitive composite materials comprised straw and mud in the form of bricks for building construction; by being forced to make bricks without straw. The ancient brick-making process can still be seen on Egyptian tomb paintings in the Metropolitan Museum of Art. The most advanced examples perform routinely on spacecraft in demanding environments. The most visible applications pave our roadways in the form of either steel and aggregate reinforced Portland cement or asphalt concrete. Those composites closest to our personal hygiene form our shower stalls and bath tubs made of fiberglass. Solid surface, imitation granite and cultured marble sinks and counter tops are widely used to enhance our living experiences.

Plywood is a common composite material many people encounter in their everyday life.

There are two categories of constituent materials: matrix and reinforcement. At least one portion of each type is required. The matrix material surrounds and supports the reinforcement materials by maintaining their relative positions. The reinforcements impart their special mechanical and physical properties to enhance the matrix properties. A synergism produces material properties unavailable from the individual constituent materials, while the wide variety of matrix and strengthening materials allow the designer of the product or structure to choose an optimum combination.

Engineered composite materials must be formed to shape. The matrix material can be introduced to the reinforcement before or after the reinforcement material is placed into the mold cavity or onto the mold surface. The matrix material experiences a melding event, after which the part shape is essentially set. Depending upon the nature of the matrix material, this melding event can occur in various ways such as chemical polymerization or solidification from the melted state.

A variety of molding methods can be used according to the end-item design requirements. The principal factors impacting the methodology are the natures of the chosen matrix and reinforcement materials. Another important factor is the gross quantity of material to be produced. Large quantities can be used to justify high capital expenditures for rapid and automated manufacturing technology. Small production quantities are accommodated with lower capital expenditures but higher labor and tooling costs at a correspondingly slower rate.

Most commercially produced composites use a polymer matrix material often called a resin solution. There are many different polymers available depending upon the starting raw ingredients. There are several broad categories, each with numerous variations. The most common are known as polyester, vinyl ester, epoxy, phenolic, polyimide, polyamide, polypropylene, PEEK, and others. The reinforcement materials are often fibers but also commonly ground minerals.

Molding Methods

In general, the reinforcing and matrix materials are combined, compacted and processed to undergo a melding event. After the melding event, the part shape is essentially set, although it can deform under certain process conditions. For a thermoset polymeric matrix material, the melding event is a curing reaction that is initiated by the application of additional heat or chemical reactivity such as organic peroxide. For a thermoplastic polymeric matrix material, the melding event is solidification from the melted state. For a metal matrix material such as titanium foil, the melding event is a fusing at high pressure and a temperature near the melt point.

For many molding methods, it is convenient to refer to one mold piece as a "lower" mold and another mold piece as an "upper" mold. Lower and upper refer to the different faces of the molded panel, not the mold's configuration in space. In this convention, there is always a lower mold, and sometimes an upper mold. Part construction begins by applying materials to the lower mold. Lower mold and upper mold are more generalized descriptors than more common and specific terms such as male side, female side, a-side, b-side, tool side, bowl, hat, mandrel, etc. Continuous manufacturing processes use a different nomenclature.

The molded product is often referred to as a panel. For certain geometries and material combinations, it can be referred to as a casting. For certain continuous processes, it can be referred to as a profile.

Open Molding

A process using a rigid, one sided mold which shapes only one surface of the panel. The opposite surface is determined by the amount of material placed upon the lower mold. Reinforcement materials can be placed manually or robotically. They include continuous fiber forms fashioned into textile constructions and chopped fiber. The matrix is generally a resin, and can be applied with a pressure roller, a spray device or manually. This process is generally done at ambient temperature and atmospheric pressure. Two variations of open molding are Hand Layup and Spray-up.

Vacuum Bag Molding

A process using a two-sided mold set that shapes both surfaces of the panel. On the lower side is a rigid mold and on the upper side is a flexible membrane. The flexible membrane can be a reusable silicone material or an extruded polymer film such as nylon. Reinforcement materials can be placed on the lower mold manually or robotically, generally as continuous fiber forms fashioned into textile constructions. The matrix is generally a resin. The fiber form may be pre-impregnated with the resin in the form of prepreg fabrics or unidirectional tapes. Otherwise, liquid matrix material is introduced to dry fiber forms prior to applying the flexible film. Then, vacuum is applied to the mold cavity. This process can be performed at either ambient or elevated temperature with ambient atmospheric pressure acting upon the vacuum bag. Most economical way is using a venturi vacuum and air compressor or a vacuum pump.

Autoclave Molding

A process using a two-sided mold set that forms both surfaces of the panel. On the lower side is a rigid mold and on the upper side is a flexible membrane made from silicone or an extruded polymer film such as nylon. Reinforcement materials can be placed manually or robotically. They include continuous fiber forms fashioned into textile constructions. Most often, they are pre-impregnated with the resin in the form of prepreg fabrics or unidirectional tapes. In some instances, a resin film is placed upon the lower mold and dry reinforcement is placed above. The upper mold is installed and vacuum is applied to the mold cavity. Then, the assembly is placed into an autoclave pressure vessel. This process is generally performed at both elevated pressure and elevated temperature. The use of elevated pressure facilitates a high fiber volume fraction and low void content for maximum structural efficiency.

Resin Transfer Molding

A process using a two-sided mold set that forms both surfaces of the panel. The lower side is a rigid mold. The upper side can be a rigid or flexible mold. Flexible molds can be made from composite materials, silicone or extruded polymer films such as nylon. The two sides fit together to produce a mold cavity. The distinguishing feature of resin transfer molding is that the reinforcement materials are placed into this cavity and the mold set is closed prior to the introduction of matrix material. Resin transfer molding includes numerous varieties which differ in the mechanics of how the resin is introduced to the reinforcement in the mold cavity. These variations include everything from vacuum infusion to vacuum assisted resin transfer molding. This process can be performed at either ambient or elevated temperature.

Other types of molding include press molding, transfer molding, pultrusion molding, filament winding, casting, centrifugal casting and continuous casting.

Tooling

Some types of tooling materials used in the manufacturing of composites structures include invar, steel, aluminum, reinforced silicon rubber, nickle, and carbon fiber. Selection of the tooling material is typically based on, but not limited to, the coefficient of thermal expansion, expected number of cycles, end item tolerance, desired or required surface condition, method of cure, glass transition temperature of the material being molded, molding method, matrix, cost and a variety of other considerations.

Mechanics of Composite Materials

The physical properties of composite materials are generally not isotropic in nature, but rather are typically orthotropic. For instance, the stiffness of a composite panel will often depend upon the directional orientation of the applied forces and/or moments. Panel stiffness is also dependent on the design of the panel. For instance, the fiber reinforcement and matrix used the method of panel build, thermoset versus thermoplastic, type of weave, and orientation of fiber axis to the primary force.

In contrast, isotropic materials (for example, aluminum or steel), in standard wrought forms, typically have the same stiffness regardless of the directional orientation of the applied forces and/or moments.

The relationship between forces/moments and strains/curvatures for an isotropic material can be described with the following material properties: Young's Modulus, the Shear Modulus and the Poisson's ratio, in relatively simple mathematical relationships. For the anisotropic material, it requires the mathematics of a second order tensor and can require up to 21 material property constants. For the special case of orthogonal isotropy, there are three different material property constants for each of Young's Modulus, Shear Modulus and Poisson's Ratio for a total of nine material property constants to describe the relationship between forces/moments and strains/curvatures.

Categories of Fiber Reinforced Composite Materials

Fiber reinforced composite materials can be divided into two main categories normally referred to as short fiber reinforced materials and continuous fiber reinforced materials. Continuous reinforced materials will often constitute a layered or laminated structure. The woven and continuous fiber styles are typically available in a variety of forms, being pre-impregnated with the given matrix (resin), dry, uni-directional tapes of various widths, plain weave, harness satins, braided, and stitched.

The short and long fibers are typically employed in compression molding and sheet molding operations. These come in the form of flakes, chips, and random mate (which can also be made from a continuous fiber laid in random fashion until the desired thickness of the ply/laminate is achieved).

Failure of Composites

Shock, impact, or repeated cyclic stresses can cause the laminate to separate at the interface between two layers, a condition known as delamination. Individual fibers can separate from the matrix e.g. fiber pull-out.

Composites can fail on the microscopic or macroscopic scale. Compression failures can occur at both the macro scale or at each individual reinforcing fiber in compression buckling. Tension failures can be net section failures of the part or degradation of the composite at a microscopic scale where one or more of the layers in the composite fail in tension of the matrix or failure the bond between the matrix and fibers.

Some composites are brittle and have little reserve strength beyond the initial onset of failure while others may have large deformations and have reserve energy absorbing capacity past the onset of damage. The variations in fibers and matrices that are available and the mixtures that can be made with blends leave a very broad range of properties that can be designed into a composite structure.

Examples of Composite Materials

Fiber Reinforced Polymers or FRPs include Wood comprising (cellulose fibers in a lignin and hemicellulose matrix), Carbon-fiber reinforced plastic or CFRP, Glass-fiber reinforced plastic or GFRP (also GRP). If classified by matrix then there are Thermoplastic Composites, short fiber thermoplastics, long fiber thermoplastics or long fiber reinforced thermoplastics There are numerous thermoset composites, but advanced systems usually incorporate aramid fiber and carbon fiber in an epoxy resin matrix.

Composites can also utilize metal fibers reinforcing other metals, as in Metal matrix composites or MMC. Ceramic matrix composites include Bone (hydroxyapatite reinforced with collagen fibers), Cermet (ceramic and metal) and Concrete. Organic matrix/ceramic aggregate composites include Asphalt concrete, Mastic asphalt, Mastic roller hybrid, Dental composite, Syntactic foam and Mother of Pearl. Chobham armour is a special composite used in military applications.

Additionally, thermoplastic composite materials can be formulated with specific metal powders resulting in materials with a density range from two g/cc to 11 g/cc (same density as lead). These materials can be used in place of traditional materials such as aluminum, stainless steel, brass, bronze, copper, lead, and even tungsten in weighting, balancing, vibration dampening, and radiation shielding applications. High density composites are an economically viable option when certain materials are deemed hazardous and are banned (such as lead) or when a secondary operation costs (such as machining, finishing, or coating) are a factor.

Engineered wood includes a wide variety of different products such as Plywood, Oriented strand board, Wood plastic composite (recycled wood fiber in polyethylene matrix), Pykrete (sawdust in ice matrix), Plastic-impregnated or laminated paper or textiles, Arborite, Formica (plastic) and Micarta. Other engineered laminate composites, such as Mallite, use a central core of end grain balsa wood, bonded to surface skins of light alloy or GRP. These generate low-weight, high rigidity materials.

Typical Products

Composite materials have gained popularity (despite their generally high cost) in high-performance products such as aerospace components (tails, wings, fuselages, propellors), boat and scull hulls, and racing car bodies. More mundane uses include fishing rods and storage tanks.

Polymer

Polymers are materials that consist of molecules formed by long chains of repeating units. They may be natural or synthetic. Many useful engineering materials are polymers, such as plastics, rubbers, fibers, adhesives, and coatings. Polymers are classified as thermoplastic polymers, thermosetting polymers (thermosets), and elastomers.

Examples of Polymers

Polymers may be divided into two categories. Natural polymers (also called biopolymers) include silk, rubber, cellulose, wool, amber, keratin, collagen, starch, DNA, and shellac. Biopolymers serve key functions in organisms, acting as structural proteins, functional proteins, nucleic acids, structural polysaccharides, and energy storage molecules.

Synthetic polymers are prepared by a chemical reaction, often in a lab. Examples of synthetic polymers include PVC (polyvinyl chloride), polystyrene, synthetic rubber, silicone, polyethylene, neoprene, and nylon. Synthetic polymers are used to make plastics, adhesives, paints, mechanical parts, and many common objects.

Synthetic polymers may be grouped into two categories. Thermoset plastics are made from a liquid or soft solid substance that can be irreversibly changed into an insoluble polymer by curing using heat or radiation. Thermoset plastics tend to be rigid and have high molecular weights. The plastic stays out of shape when deformed and typically decompose before they melt. Examples of thermoset plastics include epoxy, polyester, acrylic resins, polyurethanes, and vinyl esters. Bakelite, Kevlar, and vulcanized rubber are also thermoset plastics.

Thermoplastic polymers or thermo softening plastics are the other type of synthetic polymers. While thermoset plastics are rigid, thermoplastic polymers are solid when cool, but are pliable and can be molded above a certain temperature. While thermoset plastics form irreversible chemical bonds when cured, the bonding in thermoplastics weakens with temperature. Unlike thermosets, which decompose rather than melt, thermoplastics melt into a liquid upon heating. Examples of thermoplastics include acrylic, nylon, Teflon, polypropylene, polycarbonate, ABS, and polyethylene.

Uses of Polymers

Polymers are used in almost every area of modern living. Grocery bags, soda and water bottles, textile fibers, phones, computers, food packaging, auto parts, and toys all contain polymers.

Even more-sophisticated technology uses polymers. For example, “the membranes for water desalination, carriers used in controlled drug release and biopolymers for tissue engineering all use polymers,” according to the ACS.

Popular polymers for manufacturing include polyethylene and polypropylene. Their molecules can consist of 10,000 to 200,000 monomers.

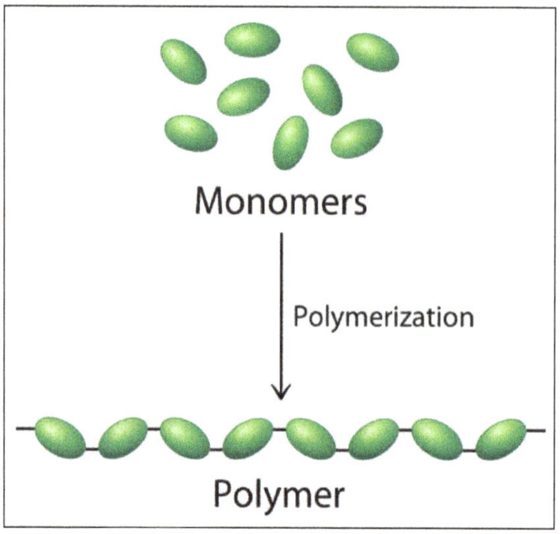

During a polymerization reaction, a large number of monomers become connected by covalent bonds to form a single long molecule, a polymer.

Thermoplastics

A thermoplastic is a type of plastic made up of polymer resins that becomes a soft material when it is heated and becomes hard when it is cooled. These materials are easily recycled and do not show any chemical property changes when they are heated or cooled multiple times. When thermoplastics are heated their physical property changes and they become a homogenized liquid and can be reshaped and resized.

In harsh acidic environments such as in cities, steel piping systems are often susceptible to rust or corrosion at the joints and therefore need special arrangements for corrosion protection. To minimize the costs associated with protecting steel pipelines that pass through severe environments, thermoplastics are considered a favorable substitute material. Some of the properties of thermoplastics that make them a suitable substitute material are:

- They handle corrosive materials and corrosive environments well.
- They are able to carry materials of extreme temperatures (either hot or cold), therefore they are good for just about every type of fluid transport application.

Thermoplastic pipes usually have a small diameter, ranging from ¼ inch to 16 inches and are often used for water pipes in residential and commercial buildings.

Common materials used to manufacture these pipes are PVC or CPVC. Additional materials include polypropylene, PVDF, ABS, nylon and polyethylene. Polyethylene gas tanks are used to transport natural gas for residential and commercial purposes.

Thermosets

A thermosetting plastic/polymer is also known a thermoset. A thermosetting plastic is a polymer that irreversibly becomes rigid when heated. Such a material is also known as a thermoset or thermosetting polymer. Initially, the polymer is a liquid or soft solid. Heat provides energy for chemical reactions that increase the cross-linking between polymer chains, curing the plastic. The rate of curing may be increasing in many cases by increasing pressure or by adding a catalyst.

Examples:

Many common plastics are thermosets. They include:

- Vulcanized rubber
- Fiberglass (a fiber-reinforced polymer composite)
- Polyester resin
- Polyurethane
- Melamine
- Bakelite
- Silicone resin
- Epoxy resin

Elastomers

An elastomer is a polymer with the property of elasticity. In other words, it is a polymer that deforms under stress and returns to its original shape when the stress is removed. The term is a contraction of the words "elastic polymer." There are many types of elastomers, most of which are rubbers. The term elastomer is therefore often used interchangeably with the term rubber. Other elastomers, which melt when heated, are classified as thermoplastic.

Rubber, an elastomer, is widely used for the manufacture of tires.

Rubbers (both natural and synthetic) are widely used for the manufacture of tires, tubes, hoses, belts, matting, gloves, toy balloons, rubber bands, pencil erasers, and adhesives. Thermoplastic elastomers are used in manufacturing processes, such as by injection molding. Thermoplastic polyurethanes are used for various applications, including the production of foam seating, seals, gaskets, and carpet underlay.

Chips of thermoplastic polyurethane.

Properties

Elastomers are amorphous polymers with considerable segmental motion. Their general molecular form has been likened to a "spaghetti and meatball" structure, where the meatballs signify cross-links between the flexible polymer chains, which are like spaghetti strands. Each polymer chain is made up of many monomer subunits, and each monomer is usually made of carbon, hydrogen, and oxygen atoms, and occasionally silicon atoms.

Most elastomers are thermosets—that is, they require curing (by heat, chemical reaction, or irradiation). In the curing process, the long polymer chains become cross-linked by covalent bonds, the material becomes stronger, and it cannot be remelted and remolded. Some elastomers are thermoplastic, melting to a liquid state when heated and turning brittle when cooled sufficiently. In thermoplastic elastomers, the polymer chains are cross-linked by weaker bonds, such as hydrogen bonds or dipole-dipole interactions.

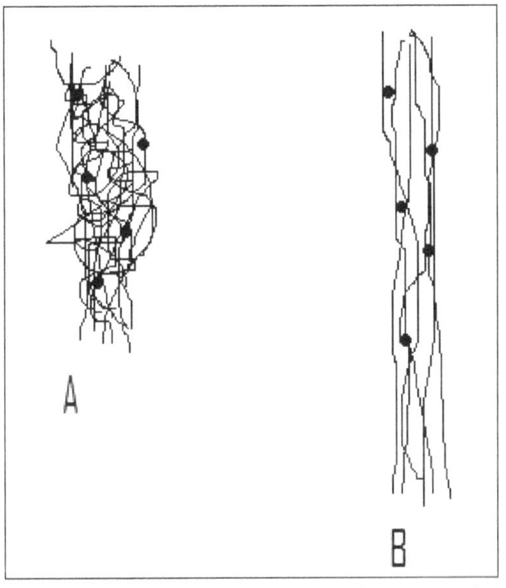

A is a schematic drawing of an unstressed polymer, and B is the same polymer under stress. When the stress is removed, B returns to the A configuration. (The dots represent cross-links.)

The elasticity is derived from the ability of the long chains to reconfigure them to distribute an applied stress. Covalent cross-linkages, in particular, ensure that the elastomer will return to its original configuration when the stress is removed. As a result of this extreme flexibility, elastomers can reversibly extend from 5 to 700 percent, depending on the specific material. Without the cross-linkages or with short, uneasily reconfigured chains, the applied stress would result in permanent deformation.

The temperature of the polymer also affects its elasticity. Elastomers that have been cooled to a glassy or crystalline phase will have less mobile chains, and consequently less elasticity, than those manipulated at temperatures higher than the glass transition temperature of the polymer. At ambient temperatures, rubbers are thus relatively soft (Young's modulus of about 3 MPa) and deformable.

Examples of Elastomers

Unsaturated rubbers that can be cured by sulfur vulcanization:

- Natural Rubber (NR);
- Synthetic Polyisoprene (IR);

- Butyl rubber (copolymer of isobutylene and isoprene, IIR):
    - Halogenated butyl rubbers (Chloro Butyl Rubber: CIIR; Bromo Butyl Rubber: BIIR).
- Polybutadiene (BR);
- Styrene-butadiene Rubber (copolymer of polystyrene and polybutadiene, SBR);
- Nitrile Rubber (copolymer of polybutadiene and acrylonitrile, NBR), also called Buna N rubbers:
    - Hydrogenated Nitrile Rubbers (HNBR) Therban and Zetpol.
- Chloroprene Rubber (CR), polychloroprene, Neoprene, Baypren etc.

Unsaturated rubbers can also be cured by non-sulfur vulcanization if desired.

Saturated Rubbers that cannot be cured by sulfur vulcanization:

- EPM (ethylene propylene rubber, a copolymer of ethylene and propylene) and EPDM rubber (ethylene propylene diene rubber a terpolymer of ethylene, propylene and a diene-component);
- Epichlorohydrin rubber (ECO);
- Polyacrylic rubber (ACM, ABR);
- Silicone rubber (SI, Q, VMQ);
- Fluorosilicone Rubber (FVMQ);
- Fluoroelastomers (FKM, and FEPM) Viton, Tecnoflon, Fluorel, Aflas and Dai-El;
- Perfluoroelastomers (FFKM) Tecnoflon PFR, Kalrez, Chemraz, Perlast;
- Polyether Block Amides (PEBA);
- Chlorosulfonated Polyethylene (CSM), (Hypalon);
- Ethylene-vinyl acetate (EVA).

Various other types of elastomers:

- Thermoplastic elastomers (TPE), for example Elastron, etc.;
- Thermoplastic Vulcanizates (TPV), for example Santoprene TPV;
- Thermoplastic Polyurethane (TPU);
- Thermoplastic Olefins (TPO);
- The proteins resilin and elastin;
- Polysulfide Rubber.

Uses

Most elastomers are rubbers, including both natural and synthetic varieties. They are used mainly for the manufacture of tires and tubes. They are also used to produce goods such as hoses, belts, matting, gloves, toy balloons, rubber bands, pencil erasers, and adhesives. As a fiber, rubber (called "elastic") is valuable for the textile industry. Ethylene propylene rubber (EPR) is useful as insulation for high voltage cables. Thermoplastic elastomers are relatively easy to use in manufacturing processes, such as by injection molding.

Polyurethanes are widely used in high-resiliency, flexible foam seating, seals, askets, carpet underlay, Spandex fibers, and electrical potting compounds.

Mathematical Background

Using the laws of thermodynamics, stress definitions, and polymer characteristics, ideal stress behavior may be calculated using the following equation:

$$\sigma = nkT\left[\lambda_1^2 + \lambda_1^{-1}\right].$$

Where, n is the number of chain segments per unit volume, k is Boltzmann's Constant, T is temperature, and λ is distortion in the 1 direction.

These findings are accurate for values up to approximately 400 percent strain. At that point, alignment between stretched chains begins to result in crystallization from non-covalent bonding.

Although Young's Modulus does not exist for elastomers because of the nonlinear nature of the stress-strain relationship, a "secant modulus" can be found at a particular strain.

References

- Metals-Nonmetals-and-Metalloids, Periodic-Properties-of-the-Elements, Chemistry: libretexts.org, Retrieved 14 March, 2019
- Ferrous-metals: civilseek.com, Retrieved 13 June, 2019
- What-are-Nonferrous-metals, nonferrous-metals, types-of-metals: weldinghandbook.com, Retrieved 17 January, 2019
- Nonmetals-definition-and-properties: thoughtco.com, Retrieved 8 March, 2019
- A-Metalloids, Periodic-Table, Chemistry: libretexts.org, Retrieved 28 May, 2019
- Alloy: newworldencyclopedia.org, Retrieved 23 January, 2019
- Ceramics-materials-joining-and-applications: twi-global.com, Retrieved 3 July, 2019
- Ceramic-Composition-and-classification: newworldencyclopedia.org, Retrieved 11 February, 2019
- What-are-refractories: refractoriesinstitute.org, Retrieved 20 August, 2019
- Refractory, technology: britannica.com, Retrieved 9 April, 2019
- Glass: materials.unsw.edu.au, Retrieved 14 August, 2019
- Composite-material: newworldencyclopedia.org, Retrieved 28 February, 2019

- What-is-a-polymer: thoughtco.com, Retrieved 8 May, 2019
- Definition-of-polymer: thoughtco.com, Retrieved 11 January, 2019
- Polymers: livescience.com, Retrieved 29 June, 2019
- Thermoplastic: corrosionpedia.com, Retrieved 20 July, 2019
- Liquid-or-soft-solid, thermosetting-plastic: thoughtco.com, Retrieved 27 April, 2019
- Elastomer: newworldencyclopedia.org, Retrieved 7 February, 2019

3

Properties of Materials

The properties of materials are divided into several categories in order to make their study more efficient. Some of these are mechanical, chemical, electrical, thermal, optical and magnetic properties. This chapter closely examines these properties of materials to provide an extensive understanding of the subject.

Mechanical Properties of Materials

Concepts of Stress and Strain

Stress can be defined by ratio of the perpindicular force applied to a specimen divided by its original cross sectional area, formally called engineering stress.

To compare specimens of different sizes, the load is calculated per unit area, also called normalization to the area. Force divided by area is called stress. In tension and compression tests, the relevant area is that perpendicular to the force. In shear or torsion tests, the area is perpendicular to the axis of rotation. The stress is obtained by dividing the load (F) by the original area of the cross section of the specimen (A_0),

$$\sigma = \frac{F}{A_o}$$

The unit is the Megapascal = 10^6 Newtons/m^2.

There is a change in dimensions, or deformation elongation, ΔL as a result of a tensile or compressive stress. To enable comparison with specimens of different length, the elongation is also normalized, this time to the length l_o. This is called strain. So, Strain is the ratio of change in length due to deformation to the original length of the specimen, formally called engineering strain. Strain is unit less, but often units of m/m (or mm/mm) are used.

The strain used for the engineering stress-strain curve is the average linear strain, which is obtained by dividing the elongation of the gage length of the specimen, by its original length,

$$\varepsilon = \frac{I_i - I_o}{I_o} = \frac{\Delta I}{I_o}$$

The strain used for the engineering stress-strain curve is the average linear strain, which is obtained by dividing the elongation of the gage length of the specimen, by its original length.

Since both the stress and the strain are obtained by dividing the load and elongation by constant factors, the load-elongation curve will have the same shape as the engineering stress-strain curve. The two curves are frequently used interchangeably.

The shape and magnitude of the stress-strain curve of a metal will depend on its composition, heat treatment, prior history of plastic deformation, and the strain rate, temperature, and state of stress imposed during the testing. The parameters used to to describe stress-strain curve are tensile strength, yield strength or yield point, percent elongation, and reduction of area. The first two are strength parameters; the last two indicate ductility.

The general shape of the engineering stress-strain curve requires further explanation. In the elastic region stress is linearly proportional to strain. When the load exceeds a value corresponding to the yield strength, the specimen undergoes gross plastic deformation. It is permanently deformed if the load is released to zero. The stress to produce continued plastic deformation increases with increasing plastic strain, i.e., the metal strain-hardens. The volume of the specimen remains constant during plastic deformation, $A \cdot L = A_0 \cdot L_0$ and as the specimen elongates, it decreases uniformly along the gage length in cross-sectional area.

Initially the strain hardening more than compensates for this decrease in area and the engineering stress (proportional to load P) continues to rise with increasing strain. Eventually a point is reached where the decrease in specimen cross- sectional area is greater than the increase in deformation load arising from strain hardening. This condition will be reached first at some point in the specimen that is slightly weaker than the rest. All further plastic deformation is concentrated in this region, and the specimen begins to neck or thin down locally. Because the cross- sectional area now is decreasing far more rapidly than strain hardening increases the deformation load, the actual *load* required to deform the specimen falls off and the engineering stress likewise continues to decrease until fracture occurs.

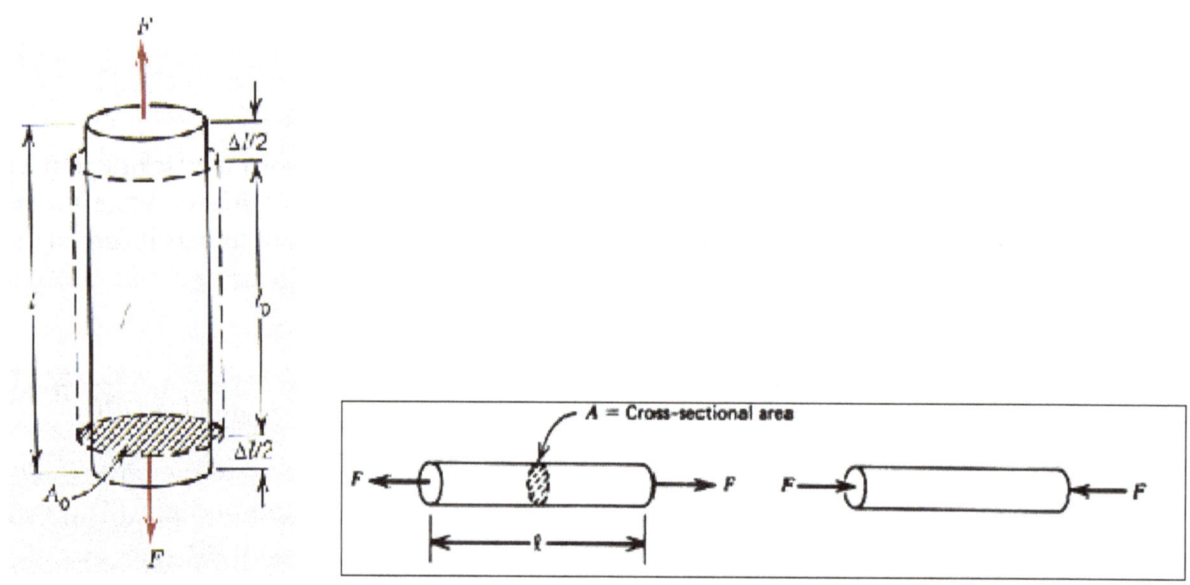

a) Tensile stress applied to a rod and b) Compressional stress applied to a rod.
Tensile and compressional stress can be defined in terms of forces applied to a uniform rod.

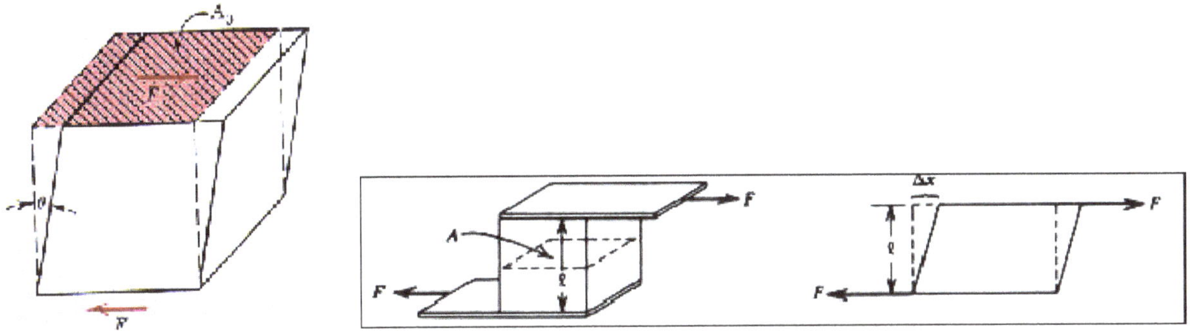

a) Shear stress results from a force couple and b) Shear stress tends to deform an object as shown. Shear stress is defined in terms of a couple that tends to deform a joining member.

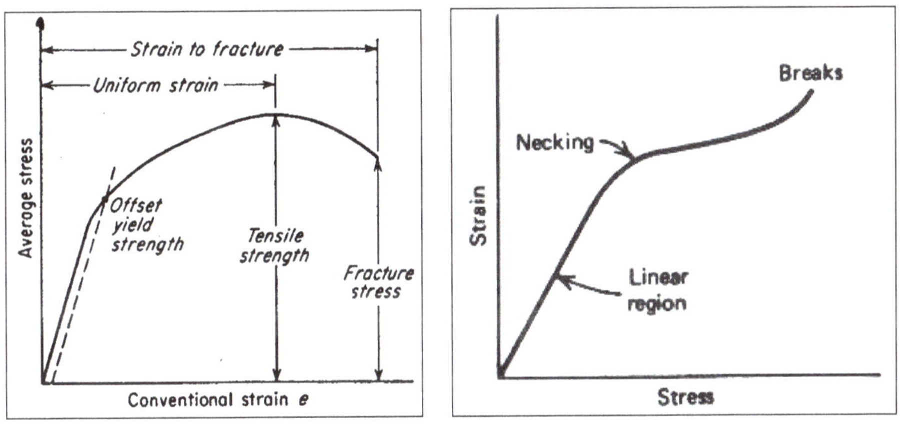

A typical stress-strain curve showing the linear region, necking and eventual break.

Shear strain is defined as the tangent of the angle theta, and, in essence, determines to what extent the plane was displaced. In this case, the force is applied as a couple (that is, not along the same line), tending to shear off the solid object that separates the force arms.

$$\text{shear strain} = \frac{\Delta x}{I}$$

where,

- *Dx*= deformation in m.
- *l* = width of a sample in m.

In this case, the force is applied as a *couple* (that is, *not* along the same line), tending to shear off the solid object that separates the force arms. In this case, the stress is again. The strain in this case is denned as the fractional change in dimension of the sheared member.

Stress-Strain Behavior

Hooke's Law

- For materials stressed in tension, at relatively low levels, stress and strain are proportional through:

  $\sigma = E\varepsilon$

- Constant E is known as the modulus of elasticity, or Young's modulus.
    - Measured in MPa and can range in values from ~4.5 × 104 - 40 × 107 MPa.

The engineering stress strain graph shows that the relationship between stress and strain is linear over some range of stress. If the stress is kept within the linear region, the material is essentially elastic in that if the stress is removed, the deformation is also gone. But if the elastic limit is exceeded, permanent deformation results. The material may begin to "neck" at some location and finally break. Within the linear region, a specific type of material will always follow the same curves despite different physical dimensions. Thus, it can say that the linearity and slope are a constant of the type of material only. In tensile and compressional stress, this constant is called the modulus of elasticity or Young's modulus (E).

$$E = \frac{F / A}{\Delta l / l}$$

where,

- Stress = F/A in N/m^2
- Strain = Dl/l unitless
- E = Modulus of elasticity in N/m^2

The modulus of elasticity has units of stress, that is, N/m^2. The following table gives the modulus of elasticity for several materials. In an exactly similar fashion, the shear modulus is defined for shear stress-strain as modulus of elasticity.

Material	Modulus (N/m^2)
Aluminum	6.89 x 10^{10}
Copper	11.73 × 10'° 20.70 × 10^{10}
Steel	2.1 x 10^8

Sress-strain Curve

The stress-strain curve characterizes the behavior of the material tested. It is most often plotted using engineering stress and strain measures, because the reference length and cross-sectional area are easily measured. Stress-strain curves generated from tensile test results help engineers gain insight into the constitutive relationship between stress and strain for a particular material.

In addition to providing quantitative information that is useful for the constitutive relationship, the stress-strain curve can also be used to qualitatively describe and classify the material. Typical regions that can be observed in a stress-strain curve are:

1. Elastic region
2. Yielding
3. Strain Hardening
4. Necking and Failure

A stress-strain curve with each region identified is shown below. The curve has been sketched using the assumption that the strain in the specimen is monotonically increasing - no unloading occurs. It should also be emphasized that a lot of variation from what's shown is possible with real materials, and each of the above regions will not always be so clearly delineated. It should be emphasized that the extent of each region in stress-strain space is material dependent, and that not all materials exhibit all of the above regions.

A stress-strain curve is a graph derived from measuring load (stress - σ) versus extension (strain - ε) for a sample of a material. The nature of the curve varies from material to material. The following diagrams illustrate the stress-strain behaviour of typical materials in terms of the engineering stress and engineering strain where the stress and strain are calculated based on the original dimensions of the sample and not the instantaneous values. In each case the samples are loaded in tension although in many cases similar behaviour is observed in compression.

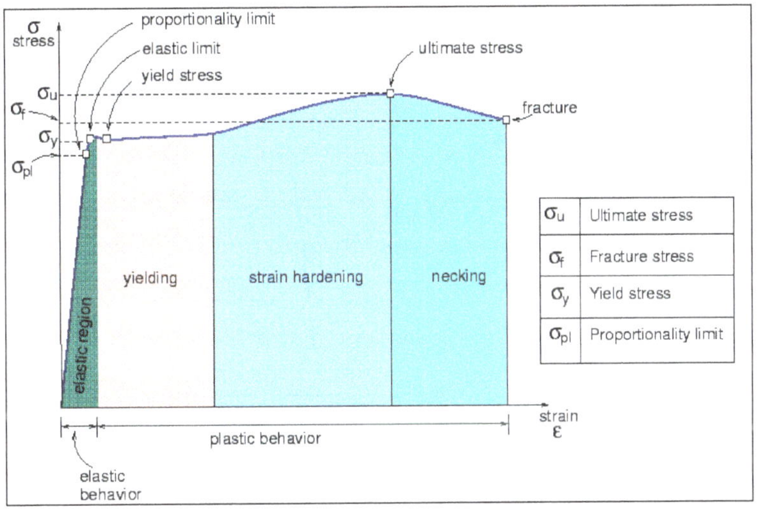

Various regions and points on the stress-strain curve.

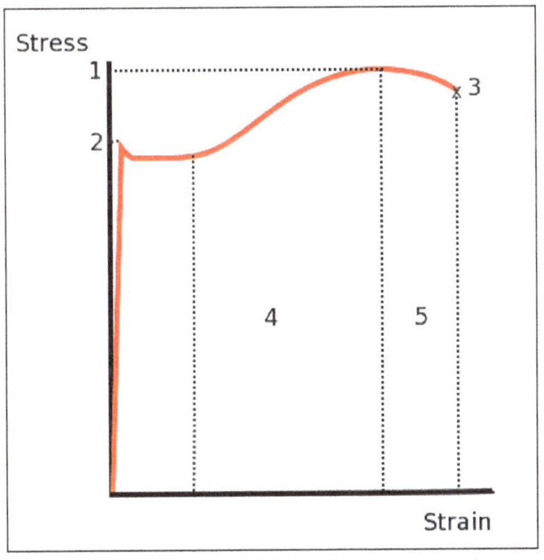

Stress vs. Strain curve for mild steel steel (Ductile material).

In figure, reference numbers are:

1. Ultimate strength

2. Yeild Strength
3. Ruputure
4. Strain hardenining region
5. Necking region

Brittle and Ductile Behavior

The behavior of materials can be broadly classified into two categories; brittle and ductile. Steel and aluminum usually fall in the class of ductile materials. Glass, ceramics, plain concrete and cast iron fall in the class of brittle materials. The two categories can be distinguished by comparing the stress-strain curves, such as the ones shown in figure.

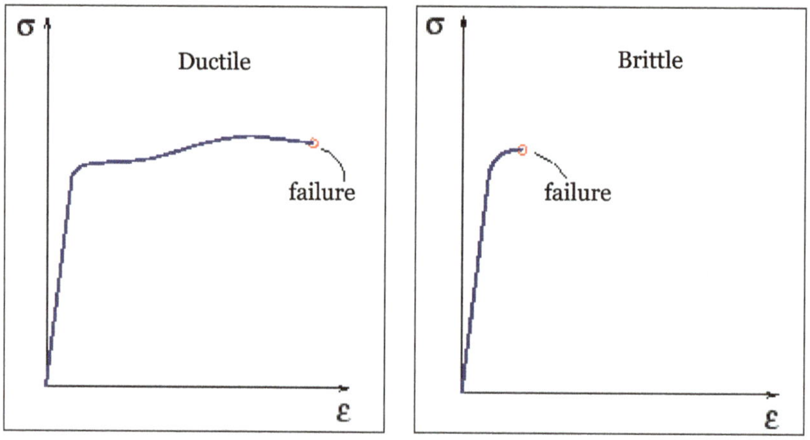

Ductile and brittle material behavior.

The material response for ductile and brittle materials are exhibited by both qualitative and quantitative differences in their respective stress-strain curves. Ductile materials will withstand large strains before the specimen ruptures; brittle materials fracture at much lower strains. The yielding region for ductile materials often takes up the majority of the stress-strain curve, whereas for brittle materials it is nearly nonexistent. Brittle materials often have relatively large Young's moduli and ultimate stresses in comparison to ductile materials.

These differences are a major consideration for design. Ductile materials exhibit large strains and yielding before they fail. On the contrary, brittle materials fail suddenly and without much warning. Thus ductile materials such as steel are a natural choice for structural members in buildings as we desire considerable warning to be provided before a building fails. The energy absorbed (per unit volume) in the tensile test is simply the area under the stress strain curve. Clearly, by comparing the curves in figure. It can be observed that ductile materials are capable of absorbing much larger quantities of energy before failure.

Finally, it should be emphasized that not all materials can be easily classified as either ductile or brittle. Material response also depends on the operating environment; many ductile materials become brittle as the temperature is decreased. With advances in metallurgy and composite technology, other materials are advanced combinations of ductile and brittle constituents.

Often in structural design, structural members are designed to be in service below the yield stress. The reason being that once the load exceeds the yield limit, the structural members will exhibit large deformations (imagine for instance a roof sagging) that are undesirable. Thus materials with larger yield strength are preferable.

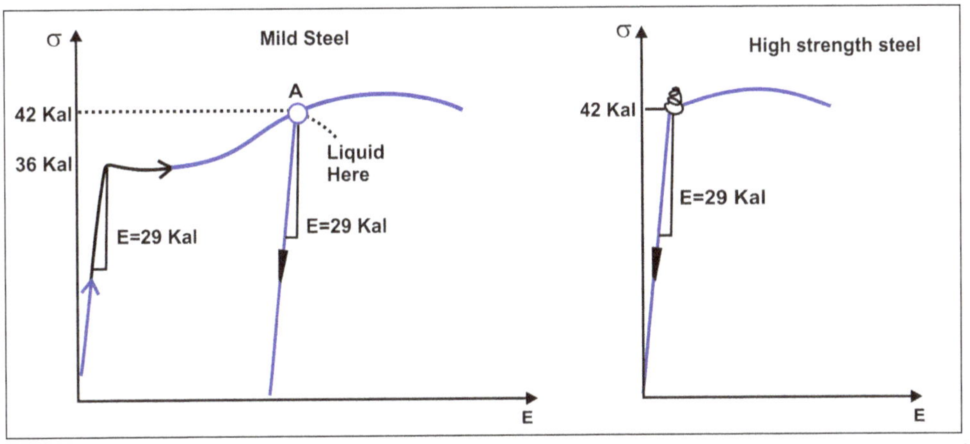

After work hardening, the stress-strain curve of a mild steel (left) resembles that of high-strength steel (right).

We will for now concentrate on steel, a commonly used structural material. Mild steels have yield strength somewhere between 240 and 360 N/mm². When work- hardened, the yield strength of this steel increases. Work hardening is the process of loading mild steel beyond its yield point and unloading as shown in figure. When the material is loaded again, the linear elastic behavior now extends up to point A as shown. The negative aspect of work hardening is some loss in ductility of the material. It is noteworthy that mild steel is usually recycled. Because of this, the yield strength may be a little higher than expected for the mild steel specimens tested in the laboratory.

Often in structural design, structural members are designed to be in service below the yield stress. The reason being that once the load exceeds the yield limit, the structural members will exhibit large deformations (imagine for instance a roof sagging) that are undesirable. Thus materials with larger yield strength are preferable.

Generally, the stress strain distribution varies from a material to another and could be in different forms as follows. Consequently, the type of material and fracture pattern can be defined and determined according to its stress-strain distribution diagram.

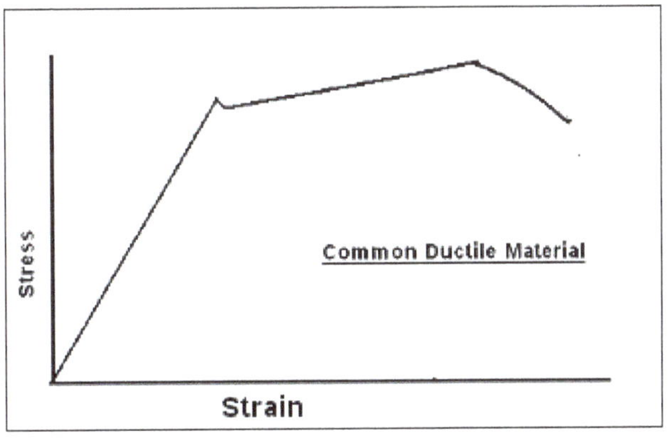

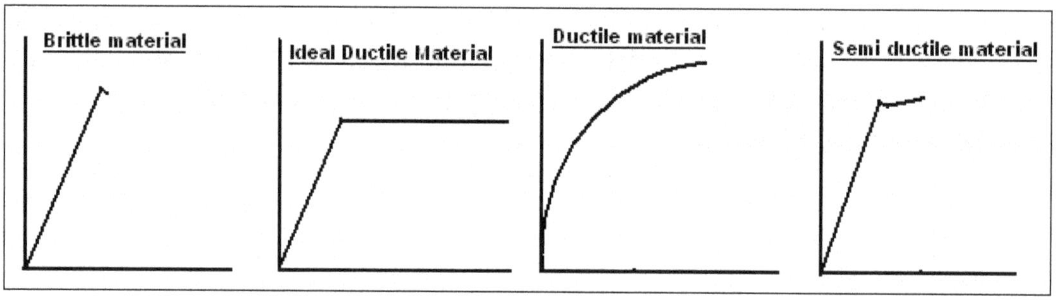

Various stress-strain diagrhams for different engineerin materials.

Yield Strength

The yield point is defined in engineering and materials science as the stress at which a material begins to plastically deform. Prior to the yield point the material will deform elastically and will return to its original shape when the applied stress is removed. Once the yield point is passed some fraction of the deformation will be permanent and non-reversible. Knowledge of the yield point is vital when designing a component since it generally represents an upper limit to the load that can be applied. It is also important for the control of many materials production techniques such as forging, rolling, or pressing.

In structural engineering, yield is the permanent plastic deformation of a structural member under stress. This is a soft failure mode which does not normally cause catastrophic failure unless it accelerates buckling.

It is often difficult to precisely define yield due to the wide variety of stress-strain behaviors exhibited by real materials. In addition there are several possible ways to define the yield point in a given material.

Yield occurs when dislocations first begin to move. Given that dislocations begin to move at very low stresses, and the difficulty in detecting such movement, this definition is rarely used.

- Elastic Limit - The lowest stress at which permanent deformation can be measured. This requires a complex interactive load-unload procedure and is critically dependent on the accuracy of the equipment and the skill of the operator.
- Proportional Limit - The point at which the stress-strain curve becomes non- linear. In most metallic materials the elastic limit and proportional limit are essentially the same.
- Offset Yield Point (proof stress) - Due to the lack of a clear border between the elastic and plastic regions in many materials, the yield point is often defined as the stress at some arbitrary plastic strain (typically 0.2%). This is determined by the intersection of a line offset from the linear region by the required strain. In some materials there is essentially no linear region and so a certain value of plastic strain is defined instead. Although somewhat arbitrary this method does allow for a consistent comparison of materials and is the most common.

Yield Point

If the stress is too large, the strain deviates from being proportional to the stress. The point at which this happens is the *yield point* because there the material yields, deforming permanently (plastically).

Yield stress: Hooke's law is not valid beyond the yield point. The stress at the yield point is called *yield stress*, and is an important measure of the mechanical properties of materials. In practice, the yield stress is chosen as that causing a permanent strain of 0.002, which called as proof stress. The yield stress measures the resistance to plastic deformation.

The *yield strength* is the stress required to produce a small-specified amount of plastic deformation. The usual definition of this property is the *offset yield strength* determined by the stress corresponding to the intersection of the stress- strain curve and a line parallel to the elastic part of the curve offset by a specified strain. In the United States the offset is usually specified as a strain of 0.2 or 0.1 percent (e = 0.002 or 0.001).

$$R_{p0.2} = \frac{P_{(strain\ offset = 0.002)}}{A_0}$$

A good way of looking at offset yield strength is that after a specimen has been loaded to its 0.2 percent offset yield strength and then unloaded it will be 0.2 percent longer than before the test. The offset yield strength is often referred to in Great Britain as the *proof stress*, where offset values are either 0.1 or 0.5 percent. The yield strength obtained by an offset method is commonly used for design and specification purposes because it avoids the practical difficulties of measuring the elastic limit or proportional limit.

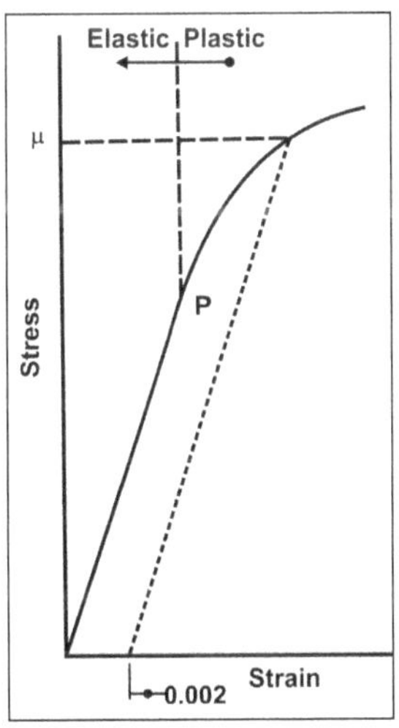

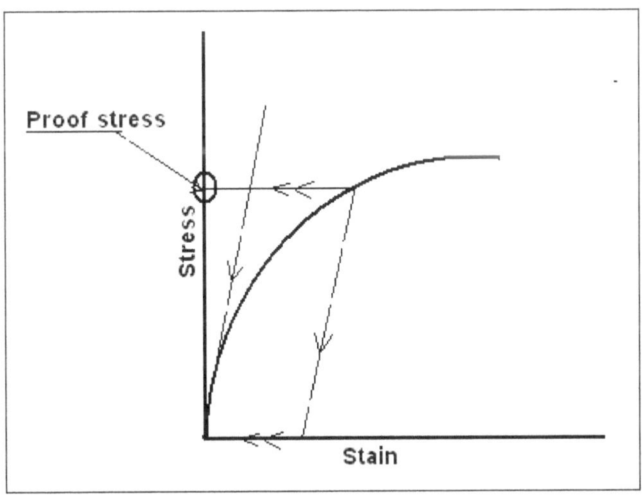

Determination of proof stress

Some materials have essentially no linear portion. to their stress-strain curve, for example, soft copper or gray cast iron. For these materials the offset method cannot be used and the usual practice is to define the yield strength as the stress to produce some total strain, for example, **e** = 0.005.

Determination of Yield Strength in Ductile Materials

In many materials, the yield stress is not very well defined and for this reason a standard has been developed to determine its value. The standard procedure is to project a line parallel to the initial elastic region starting at 0.002 strains. The 0.002 strain point is often referred to as the 2% offset strain point. The intersection of this new line with the stress-strain curve then defines the *yield strength* as shown in figure.

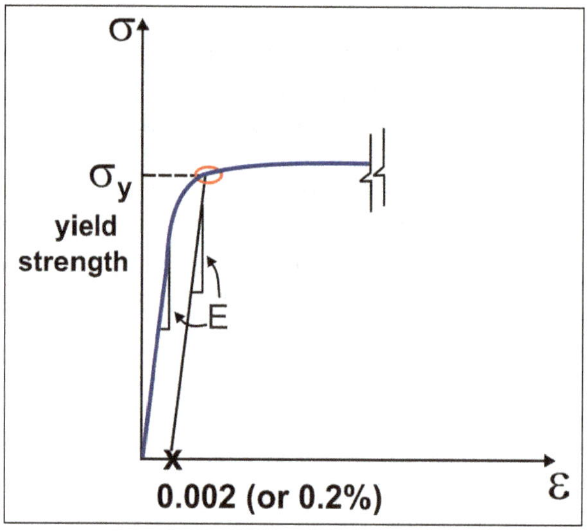

Elastic Properties of Materials

When the stress is removed, the material returns to the dimension it had before the load was applied. Valid for small strains (except the case of rubbers).

Deformation is Reversible, Non-permanent

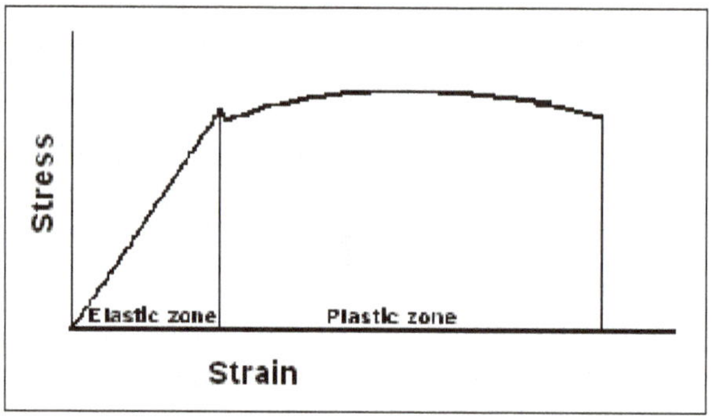

Materials subject to tension shrink laterally. Those subject to compression, bulge. The ratio of lateral and axial strains is called the *Poisson's ratio.*

When a material is placed under a tensile stress, an accompanying strain is created in the same direction.

Poisson's ratio is the ratio of the lateral to axial strains.

$$v = -\frac{\varepsilon_x}{\varepsilon_z} = \frac{\varepsilon_y}{\varepsilon_z}$$

The elastic modulus, shear modulus and Poisson's ratio are related by $E = 2G(1 + v)$

$$E = 2G(I + v)$$

- Theoretically, isotropic materials will have a value for Poisson's ratio of 0.25.

- The maximum value of υ is 0.5.
- Most metals exhibit values between 0.25 and 0.35.

Plastic Deformation

When the stress is removed, the material does not return to its previous dimension but there is a permanent, irreversible deformation. For metallic materials, elastic deformation only occurs to strains of about 0.005. After this point, plastic (non-recoverable) deformation occurs, and Hooke's Law is no longer valid. On an atomic level, plastic deformation is caused by slip, where atomic bonds are broken by dislocation motion, and new bonds are formed.

Anelasticity

Here the behavior is elastic but not the stress-strain curve is not immediately reversible. It takes a while for the strain to return to zero. The effect is normally small for metals but can be significant for polymers.

Tensile Strength

When stress continues in the plastic regime, the stress-strain passes through a maximum, called the tensile strength (s_{TS}), and then falls as the material starts to develop a neck and it finally breaks at the fracture point. Note that it is called strength, not stress, but the units are the same, MPa.

For structural applications, the yield stress is usually a more important property than the tensile strength, since once it is passed, the structure has deformed beyond acceptable limits.

The tensile strength, or ultimate tensile strength (UTS), is the maximum load divided by the original cross-sectional area of the specimen.

$$R_m = \frac{P_{max}}{A_0}$$

The tensile strength is the value most often quoted from the results of a tension test; yet in reality it is a value of little fundamental significance with regard to the strength of a metal. For ductile metals the tensile strength should be regarded as a measure of the maximum load, which a metal can withstand under the very restrictive conditions of uniaxial loading. It will be shown that this value bears little relation to the useful strength of the metal under the more complex conditions of stress, which are usually encountered.

For many years it was customary to base the strength of members on the tensile strength, suitably reduced by a factor of safety. The current trend is to the more rational approach of basing the static design of ductile metals on the yield strength.

However, because of the long practice of using the tensile strength to determine the strength of materials, it has become a very familiar property, and as such it is a very useful identification of a material in the same sense that the chemical composition serves to identify a metal or alloy.

Further, because the tensile strength is easy to determine and is a quite reproducible property, it is useful for the purposes of specifications and for quality control of a product. Extensive empirical correlations between tensile strength and properties such as hardness and fatigue strength are often quite useful. For brittle materials, the tensile strength is a valid criterion for design.

Ductility

The ability to deform before braking. It is the opposite of brittleness. Ductility can be given either as percent maximum elongation e_{max} or maximum area reduction.

At our present degree of understanding, ductility is a qualitative, subjective property of a material. In general, measurements of ductility are of interest in three ways:

1. To indicate the extent to which a metal can be deformed without fracture in metal working operations such as rolling and extrusion.

2. To indicate to the designer, in a general way, the ability of the metal to flow plastically before fracture. A high ductility indicates that the material is "forgiving" and likely to deform locally without fracture should the designer err in the stress calculation or the prediction of severe loads.

3. To serve as an indicator of changes in impurity level or processing conditions. Ductility measurements may be specified to assess material quality even though no direct relationship exists between the ductility measurement and performance in service.

The conventional measures of ductility that are obtained from the tension test are the engineering strain at fracture e_f (usually called the *elongation*) and the *reduction of area* at fracture q. Both of these properties are obtained after fracture by putting the specimen back together and taking measurements of L_f and A_f:

$$e_f = \frac{L_f - L_0}{L_0}$$

$$e_f = \frac{A_0 - A_f}{A_0}$$

Because an appreciable fraction of the plastic deformation will be concentrated in the necked region of the tension specimen, the value of e_f will depend on the gage length L_o over which the measurement was taken. The smaller the gage length the greater will be the contribution to the overall elongation from the necked region and the higher will be the value of e_f. Therefore, when reporting values of percentage elongation, the gage length L_o always should be given.

The reduction of area does not suffer from this difficulty. Reduction of area values can be converted into an equivalent zero-gage-length elongation e_o. From the constancy of volume relationship for plastic deformation $A*L = A_o*L_o$, we obtain:

$$\frac{L}{L_0} = \frac{A_0}{A} = \frac{1}{1-q}, \quad e_0 = \frac{L-L_0}{L_0} = \frac{A_0}{A} - 1 = \frac{1}{1-q} - 1 = \frac{q}{1-q}$$

This represents the elongation based on a very short gage length near the fracture.

Another way to avoid the complication from necking is to base the percentage elongation on the uniform strain out to the point at which necking begins. The uniform elongation e_u correlates well with stretch-forming operations. Since the engineering stress-strain curve often is quite flat in the vicinity of necking, it may be difficult to establish the strain at maximum load without ambiguity. In this case the method suggested by Nelson and Winlock is useful.

Resilience

The resilience of the material is the triangular area underneath the elastic region of the curve. Resilience generally means the ability to recover from (or to resist being affected by) some shock, insult, or disturbance. However, it is used quite differently in different fields.

In physics and engineering, resilience is defined as the capacity of a material to absorb energy when it is deformed elastically and then, upon unloading to have this energy recovered. In other words, it is the maximum energy per volume that can be elastically stored. It is represented by the area under the curve in the elastic region in the Stress-Strain diagram.

Modulus of Resilience, U_r, can be calculated using the following formula:

$$U_r = \frac{\sigma^2}{2E} = 0.5\sigma \in = 0.5\sigma\left(\frac{\sigma}{E}\right),$$

where, σ is yield stress, E is Young's modulus, and ε is strain.

The ability of a material to absorb energy when deformed elastically and to return it when unloaded is called resilience. This is usually measured by the modulus of resilience, which is the strain energy per unit volume required to stress the material from, zero stress to the yield stress. The strain energy per unit volume for uniaxial tension is:

$$U_0 = \frac{1}{2}\sigma_x e_x$$

Table gives some values of modulus of resilience for different materials.

Table: Modulus of resilience for various materials.

Material	E, psi	s_o, psi	Modulus of resilience, Ur
Medium-carbon steel	30×10^6	45000	33,7
High-carbon spring steel	30×10^6	140000	320
Duraluminium	$10{,}5\times10^6$	18000	17,0
Cooper	16×10^6	4000	5,3
Rubber	150	300	300
Acrylic polymer	$0{,}5\times10^6$	2000	4,0

Toughness

The area underneath the stress-strain curve is the toughness of the material- i.e. the energy the material can absorb prior to rupture. It also can be defined as the resistance of a material to crack propogation.

In materials science and metallurgy, toughness is the resistance to fracture of a material when stressed. It is defined as the amount of energy that a material can absorb before rupturing, and can be found by finding the area (i.e., by taking the integral) underneath the stress-strain curve.

The ability of a metal to deform plastically and to absorb energy in the process before fracture is termed toughness. The emphasis of this definition should be placed on the ability to absorb energy before fracture. Recall that ductility is a measure of how much something deforms plastically before fracture, but just because a material is ductile does not make it tough. The key to toughness is a good combination of strength and ductility. A material with high strength and high ductility will have more toughness than a material with low strength and high ductility. Therefore, one way to measure toughness is by calculating the area under the stress strain curve from a tensile test. This value is simply called "material toughness" and it has units of energy per volume. Material toughness equates to a slow absorption of energy by the material.

The toughness of a material is its ability to absorb energy in the plastic range. The ability to withstand occasional, stresses above the yield stress without fracturing is particularly desirable in parts such as freight-car couplings, gears, chains, and crane hooks. Toughness is a commonly used concept, which is difficult to pin down and define. One way of looking at toughness is to consider that it is the total area under the stress-strain curve. This area is an indication of the amount of work per unit volume, which can be done, on the material without causing it to rupture. The following figure shows the stress-strain curves for high- and low-toughness materials. The high-carbon spring steel has a higher yield strength and tensile strength than the medium-carbon structural steel.

However, the structural steel is more ductile and has a greater total elongation. The total area under the stress train curve is greater for the structural steel, and therefore it is a tougher material. This illustrates that toughness is a parameter that comprises both strength and ductility. The crosshatched regions in figure indicate the modulus of resilience for each steel. Because of its higher yield strength, the spring steel has the greater resilience.

Several mathematical approximations for the area under the stress-strain curve have been suggested. For ductile metals that have a stress-strain curve like that of the structural steel, the area under the curve can be approximated by either of the following equations:

$$U_T \approx S_U\, e_f \qquad U_T \approx \frac{S_o + Su}{2} . e_f$$

For brittle materials the stress-strain curve is sometimes assumed to be a parabola, and the area under the curve is given by:

$$U_T \approx \frac{2}{3} S_U\, e_f$$

All these relations are only approximations to the area under the stress-strain curves. Further, the curves do not represent the true behavior in the plastic range, since they are all based on the original area of the specimen.

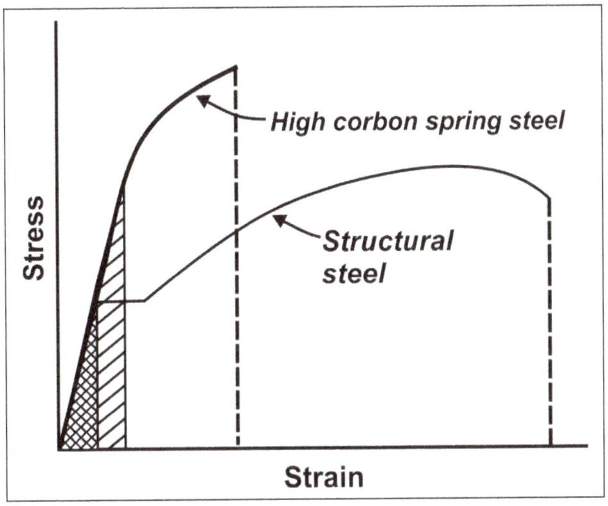

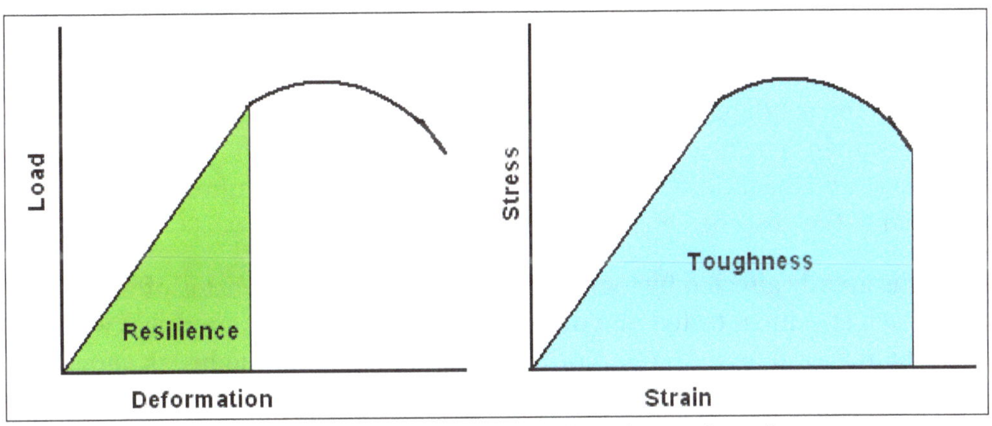

Comparison between resilience and toughness of metals.

Impact Toughness

Three of the toughness properties are:

1. Impact toughness
2. Notch toughness
3. Fracture toughness

The impact toughness (AKA Impact strength) of a material can be determined with a Charpy or Izod test. These tests are named after their inventors and were developed in the early 1900's before fracture mechanics theory was available.

Impact properties are not directly used in fracture mechanics calculations, but the economic impact tests continue to be used as a quality control method to assess notch sensitivity and for comparing the relative toughness of engineering materials.

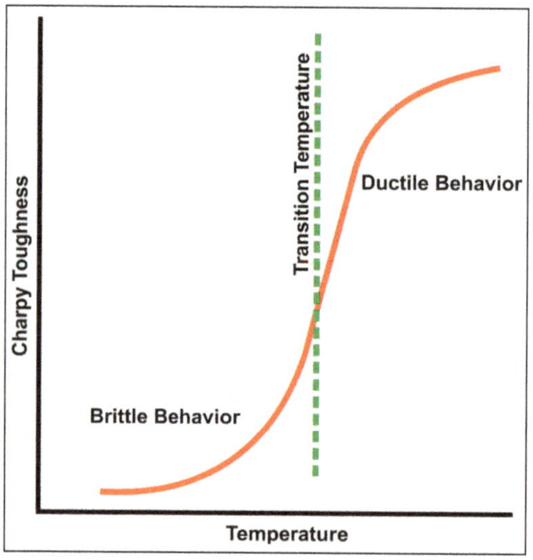

The two tests use different specimens and methods of holding the specimens, but both tests make use of a pendulum-testing machine. For both tests, the specimen is broken by a single overload event due to the impact of the pendulum. A stop pointer is used to record how far the pendulum swings back up after fracturing the specimen. The impact toughness of a metal is determined by measuring the energy absorbed in the fracture of the specimen. This is simply obtained by noting the height at which the pendulum is released and the height to which the pendulum swings after it has struck the specimen. The height of the pendulum times the weight of the pendulum produces the potential energy and the difference in potential energy of the pendulum at the start and the end of the test is equal to the absorbed energy.

Since toughness is greatly affected by temperature, a Charpy or Izod test is often repeated numerous times with each specimen tested at a different temperature. This produces a graph of impact toughness for the material as a function of temperature. Impact toughness versus temperature graph for steel is shown in the image. It can be seen that at low temperatures the material is more brittle and impact toughness is low. At high temperatures the material is more ductile and impact toughness is higher. The transition temperature is the boundary between brittle and ductile behavior and this temperature is often an extremely important consideration in the selection of a material.

Notch Toughness

Notch toughness is the ability that a material possesses to absorb energy in the presence of a flaw. As mentioned previously, in the presence of a flaw, such as a notch or crack, a material will likely exhibit a lower level of toughness. When a flaw is present in a material, loading induces a triaxial tension stress state adjacent to the flaw. The material develops plastic strains as the yield stress is exceeded in the region near the crack tip. However, the amount of plastic deformation is restricted by the surrounding material, which remains elastic. When a material is prevented from deforming plastically, it fails in a brittle manner.

Notch-toughness is measured with a variety of specimens such as the Charpy V- notch impact specimen or the dynamic tear test specimen. As with regular impact testing the tests are often repeated numerous times with specimens tested at a different temperature. With these specimens and by varying the loading speed and the temperature, it is possible to generate curves such as those shown in the graph. Typically only static and impact testing is conducted but it should be recognized that many components in service see intermediate loading rates in the range of the dashed red line.

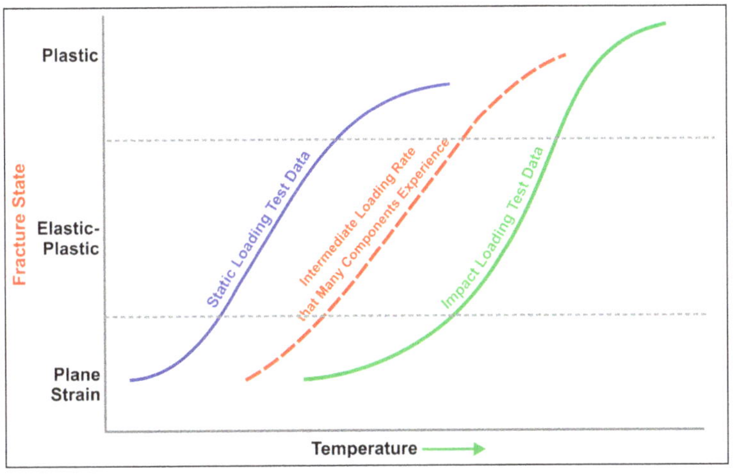

Fracture Toughness

In materials science, fracture toughness is a property which describes the ability of a material containing a crack to resist fracture, and is one of the most important properties of any material for virtually all design applications. It is denoted K_{1c} and has the units of $MPa\sqrt{m}$.

The subscript '1c' denotes mode 1 crack opening or plain strain, the material has to be too thick to shear, mode 2, or tear, mode 3.

Fracture toughness is a quantitative way of expressing a material's resistance to brittle fracture when a crack is present. If a material has a large value of fracture toughness it will probably undergo ductile fracture. Brittle fracture is very characteristic of materials with a low fracture toughness value.

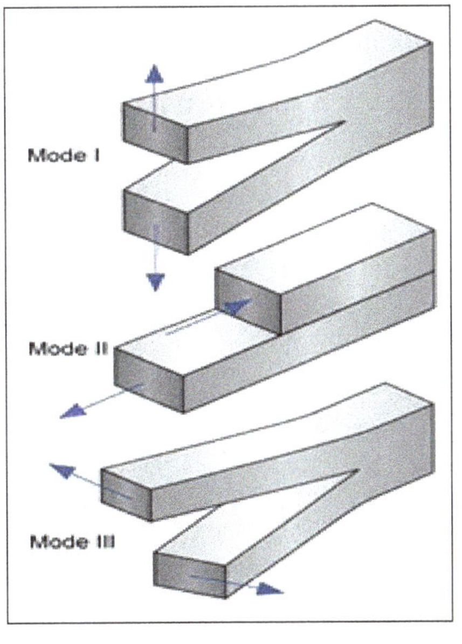

Fracture mechanics, which leads to the concept of fracture toughness, was largely based on the work of A. A. Griffith who, amongst other things, studied the behaviour of cracks in brittle materials.

Fracture toughness is an indication of the amount of stress required to propagate a pre-existing flaw. It is a very important material property since the occurrence of flaws is not completely avoidable in the processing, fabrication, or service of a material/component. Flaws may appear as cracks, voids, metallurgical inclusions, weld defects, design discontinuities, or some combination thereof. Since engineers can never be totally sure that a material is flaw free, it is common practice to assume that a flaw of some chosen size will be present in some number of components.

This approach uses the flaw size and features, component geometry, loading conditions and the material property called fracture toughness to evaluate the ability of a component containing a flaw to resist fracture.

A parameter called the stress-intensity factor (K) is used to determine the fracture toughness of most materials. A Roman numeral subscript indicates the mode of fracture and the three modes of fracture are illustrated in the image to the right. Mode I fracture is the condition in which the crack plane is normal to the direction of largest tensile loading. This is the most commonly encountered mode and, therefore, for the remainder of the material we will consider K_I.

The stress intensity factor is a function of loading, crack size, and structural geometry. The stress intensity factor may be represented by the following equation:

$$K_1 = \sigma\sqrt{\pi a \beta}$$

Where:

- K_I = Is the fracture toughness in $Mpa\sqrt{m}\left(psi\sqrt{in}\right)$.
- σ = Is the applied stress in MPa or psi;
- a:
    - Is the crack length in meters or inches;
    - Is crack length and component geometry.
- B = Factor that is different for each specimen and is dimensionless.

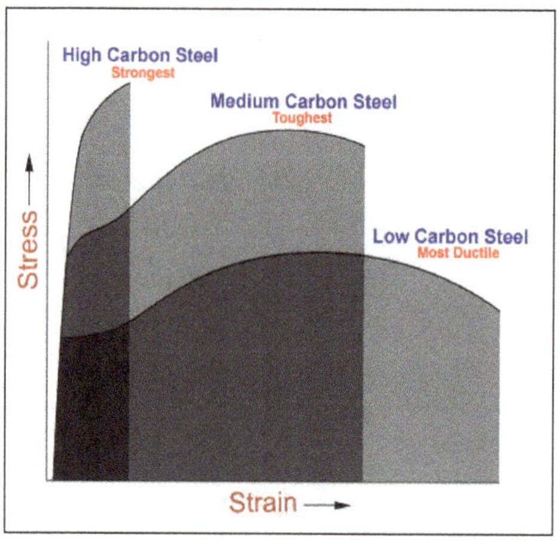

There are several variables that have a profound influence on the toughness of a material. These variables are:

- Strain rate (rate of loading)
- Temperature
- Notch effect

A metal may possess satisfactory toughness under static loads but may fail under dynamic loads or impact. As a rule ductility and, therefore, toughness decrease as the rate of loading increases. Temperature is the second variable to have a major influence on its toughness. As temperature is lowered, the ductility and toughness also decrease. The third variable is termed notch effect, has to do with the distribution of stress. A material might display good toughness when the applied stress is uniaxial; but when a multi-axial stress state is produced due to the presence of a notch, the material might not withstand the simultaneous elastic and plastic deformation in the various directions.

There are several standard types of toughness test that generate data for specific loading conditions and component design approaches.

Material Types

Brittle materials such as concrete or ceramics do not have a yield point. For these materials the rupture strength and the ultimate strength are the same.

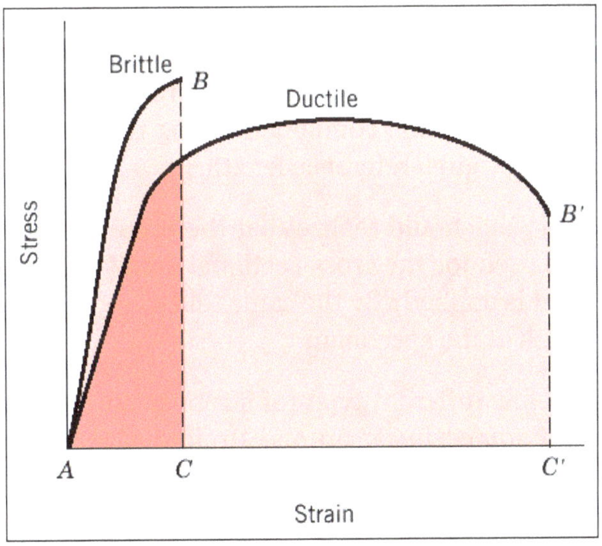

Ductile material (such as steel) generally exhibits a very linear stress-strain relationship up to a well-defined yield point. The linear portion of the curve is the elastic region and the slope is the modulus of elasticity or Young's Modulus. After the yield point the curve typically decreases slightly due to dislocations escaping from Cottrell atmospheres. As deformation continues the stress increases due to strain hardening until it reaches the ultimate strength. Until this point the cross-sectional area decreases uniformly due to Poisson contractions. However, beyond this point a *neck* forms where the local cross-sectional area decreases more quickly than the rest of the sample resulting in an increase in the true stress. On an engineering stress-strain curve this is seen as a decrease in the stress. Conversely, if the curve is plotted in terms of *true stress* and *true strain* the stress will continue to rise until failure. Eventually the neck becomes unstable and the specimen ruptures (fractures).

Most ductile metals other than steel do not have a well-defined yield point. For these materials the yield strength is typically determined by the "offset yield method", by which a line is drawn parallel to the linear elastic portion of the curve and intersecting the abscissa at some arbitrary value (most commonly. 2%). The intersection of this line and the stress-strain curve is reported as the yield point.

a. Ductile materials: Extensive plastic deformation and energy absorption "toughness") before fracture. Ductile materials can be classified into various classifications; 1- Very ductile, soft metals (e.g. Pb, Au) at room temperature, other metals, polymers, glasses at high temperature., 2- Moderately ductile fracture, typical for ductile metals, 3- Brittle fracture, cold metals, ceramics.

b. Brittle materials: Has a little plastic deformation and low energy absorption before fracture.

True Stress and Strain

When one applies a constant tensile force the material will break after reaching the tensile strength. The material starts necking (the transverse area decreases) but the stress cannot increase beyond σ_{TS}. The ratio of the force to the initial area, what we normally do, is called the engineering stress. If the ratio is to the actual area (that changes with stress) one obtains the true stress.

Stress has units of a force measure divided by the square of a length measure, and the average stress on a cross-section in the tensile test is the applied force divided by the cross-sectional area. Similarly, we may approximate the strain component along the long axis of the specimen as the change in length divided by the original reference length.

It sounds simple enough, but you should realize that there are still some choices to make. Specifically, what area should be used for the cross-sectional area? Should you use the original area or the current area as the load is applied? By the same token, should changes in length always be compared to the original length of the specimen.

The answer is that we will define different types of stress strain measures according to the way we perform the calculations. *Engineering* stress and strain measures are distinguished by the use of fixed reference quantities, typically the original cross-sectional area or original length. More precisely,

$$\sigma_E = \frac{P}{A_0}, \in_E = \frac{\Delta l}{l_o}.$$

In most engineering applications, these definitions are accurate enough, because the cross-sectional area and length of the specimen do not change substantially while loads are applied. In other situations (such as the tensile test), the cross- sectional area and the length of the specimen can change substantially. In such cases, the engineering stress calculated using the above definition (as the ratio of the applied load to the undeformed cross-sectional area) ceases to be an accurate measure. To overcome this issue alternative stress and strain measures are available.

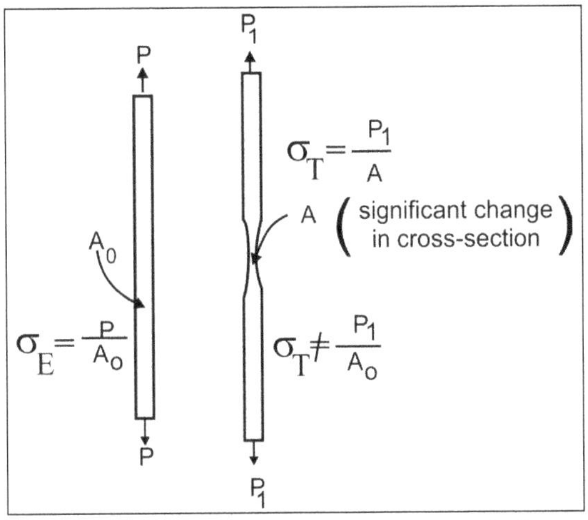

Engineering stress measures vs. true stress measures. The latter accounts for the change in cross-sectional area as the loads are applied.

1. True Stress: The true stress is defined as the ratio of the applied load (P) to the instantaneous cross-sectional area (A):

True stress can be related to the engineering stress if we assume that there is no volume change in the specimen. Under this assumption, stress can be related to the engineering stress if we assume that there is no volume change in the specimen.

$$A \cdot l = A_0 \cdot l_0,$$

which leads to:

$$\sigma_T = \frac{P}{A} = \frac{P}{A_0} \cdot \frac{l}{l_0} \sigma_E (1 + \in E).$$

2. True Strain: The true strain is defined as the sum of all the instantaneous engineering strains. Letting,

$$d\in = \frac{dl}{l}$$

the true strain is then,

$$\in_T \int d\in = \int_{l_0}^{l_f} \frac{dl}{l} \quad \in_T = \int d\in = \int_{l_0}^{l_f} \frac{dl}{l}$$

$$= \text{In} \frac{l_f}{l_0} \frac{l_f}{l_0}.$$

Where, l_f is the final length when the loading process is terminated. True strain can also be related back to the engineering strain, through the manipulation where l_f is the final length when the loading process is terminated. True strain can also be related back to the engineering strain, through the manipulation,

$$\in_T = \text{In}^{\frac{l_f}{l_0}} = \text{In}^{\frac{l_0+\Delta l}{l_0}} = \text{In}^{(1+\in_E)}$$

In closing, you should note that the true stress and strain are practically indistinguishable from the engineering stress and strain at small deformations, as shown below in figure. You should also note that as the strain becomes large and the cross-sectional area of the specimen decreases, the true stress can be much larger than the engineering stress.

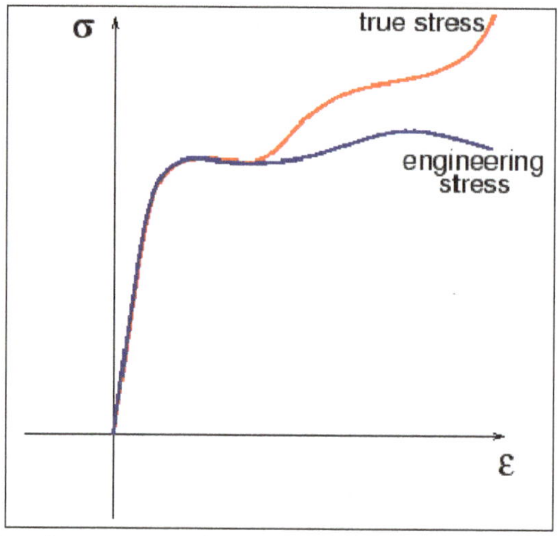

Engineering stress-strain curve vs. a true stress-strain curve.

True Stress: True Strain Curve

The engineering stress-strain curve does not give a true indication of the deformation characteristics of a metal because it is based entirely on the original dimensions of the specimen, and these dimensions change continuously during the test. Also, ductile metal which is pulled in tension becomes unstable and necks down during the course of the test. Because the cross-sectional area of the specimen is decreasing rapidly at this stage in the test, the load required continuing deformation falls off. The average stress based on original area likewise decreases, and this produces the fall-off in the stress-strain curve beyond the point of maximum load.

The engineering stress-strain curve does not give a true indication of the deformation characteristics of a metal because it is based entirely on the original dimensions of the specimen, and these dimensions change continuously during the test. Also, ductile metal which is pulled in tension becomes unstable and necks down during the course of the test. Because the cross-sectional area of the specimen is decreasing rapidly at this stage in the test, the load required continuing deformation falls off. The average stress based on original area likewise decreases, and this produces the fall-off in the stress-strain curve beyond the point of maximum load. Actually, the metal continues to strain-harden all the way up to fracture, so that the stress required to produce further deformation should also increase. If the true stress, based on the actual cross-sectional area of the specimen, is used, it is found that the stress-strain curve increases continuously up to fracture. If the strain measurement is also based on instantaneous measurements, the curve, which is obtained, is known as a true-stress-true-strain curve. This is also known as a flow curve since it represents the basic plastic-flow characteristics of the material. Any point on the flow curve can be considered the yield stress for a metal strained in tension by the amount shown on the curve. Thus, if the load is removed at this point and then reapplied, the material will behave elastically throughout the entire range of reloading. The true stress is expressed in terms of engineering stress s by:

$$\sigma = \frac{P}{A_0} \cdot (e+1) = s \cdot (e+1)$$

The derivation of this equation assumes both constancy of volume and a homogenous distribution of strain along the gage length of the tension specimen. Thus, Eq. should only be used until the onset of necking. Beyond maximum load the true stress should be determined from actual measurements of load and cross-sectional area.

$$\sigma = \frac{P}{A}$$

The true strain may be determined from the engineering or conventional strain e by:

$$å = \text{In}(e+1)$$

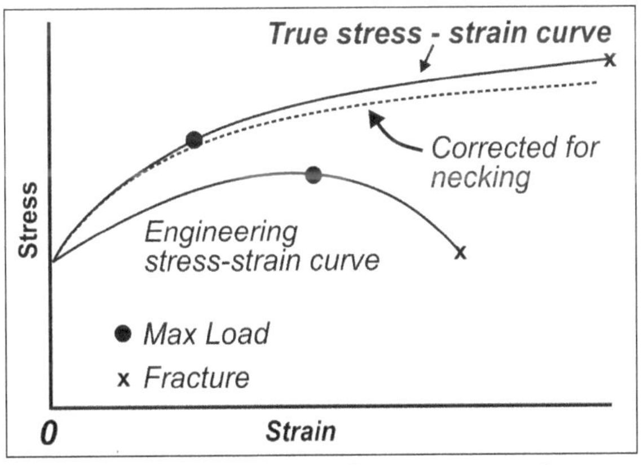

Comparison of engineering and true stress-strain curves.

Beyond maximum load the true strain should be based on actual area or diameter measurements.

$$å = \text{In}\frac{A_0}{A} = \text{In}\frac{(\pi/4)D_0^2}{(\pi/4)D^2} = 2\ln\frac{D_0}{D}$$

The above figure compares the true-stress-true-strain curve with its corresponding engineering stress-strain curve. Note that because of the relatively large plastic strains, the elastic region has been compressed into the y-axis. The true-stress-true- strain curve is always to the left of the engineering curve until the maximum load is reached. However, beyond maximum load the high-localized strains in the necked region that are used in last Equation far exceed the engineering strain calculated from first eq. Frequently the flow curve is linear from maximum load to fracture, while in other cases its slope continuously decreases up to fracture. The formation of a necked region or mild notch introduces triaxial stresses, which make it difficult to determine accurately the longitudinal tensile stress on out to fracture. The following parameters usually are determined from the true-stress-true-strain curve.

True Stress at Maximum Load

The true stress at maximum load corresponds to the true tensile strength. For most materials necking begins at maximum load at a value of strain where the true stress equals the slope of the flow curve. Let $_u$ and $_u$ denote the true stress and true strain at maximum load when the cross-sectional area of the specimen is Au. The ultimate tensile strength is given by:

$$S_u = \frac{P_{max}}{A_0}$$

$$\sigma_u = \frac{P_{max}}{A_o}$$

$$\varepsilon_u = \text{In}\frac{A_o}{A_u}$$

Eliminating P_{max} yields

$$\sigma_u = S_u \frac{A_0}{A_u}$$

$$\sigma_u = S_u e^{z_o}$$

True Fracture Stress

The true fracture stress is the load at fracture divided by the cross-sectional area at fracture. This stress should be corrected for the, triaxial state of stress existing in the tensile specimen at fracture. Since the data required for this correction are often not available, true-fracture-stress values are frequently in error.

True Fracture Strain

The true fracture strain $_f$ is the true strain based on the original area A_o and the area after fracture A_f:

$$\varepsilon_f = \ln\frac{A_0}{A_f}$$

This parameter represents the maximum true strain that the material can withstand before fracture and is analogous to the total strain to fracture of the engineering stress-strain curve. Since Equation is not valid beyond the onset of necking, it is not possible to calculate $_f$ from measured values of $_f$. However, for cylindrical tensile specimens the reduction of area q is related to the true fracture strain by the relationship:

$$\varepsilon = \ln\frac{1}{1q}$$

True Uniform Strain

The true uniform strain is the true strain based only on the strain up to maximum load. It may be calculated from either the specimen cross-sectional area A_u or the gage length L_u at maximum load. The uniform strain is often useful in estimating the formability of metals from the results of a tension test,

$$\varepsilon_u = \mathrm{In}\frac{A_0}{A_c}$$

True Local Necking Strain

The local necking strain is the strain required to deform the specimen from maximum load to fracture,

$$\varepsilon - \ln\frac{A_u}{A_f}$$

The flow curve of many metals in the region of uniform plastic deformation can be expressed by the simple power curve relation:

where n is the strain $\sigma = K.\varepsilon^n$ hardening exponent and K is the strength coefficient. A log-log plot of true stress and true strain up to maximum load will result in a straight-line. For most metals n has values between 0.10 and 0.50.

Table: Values for n and K for metals at room temperature.

Metal	Condition	n	K, psi
0,05% C steel	Annealed	0,26	77000
SAE 4340 steel	Annealed	0,15	93000
0,60% C steel	Quenched and tempered 1000 °F	0,10	228000
0,60% C steel	Quenched and tempered 1300 °F	0,19	178000
Copper	Annealed	0,54	46400
70/30 brass	Annealed	0,49	130000

Elastic Recovery During Plastic Deformation

If a material is taken beyond the yield point (it is deformed plastically) and the stress is then released, the material ends up with a permanent strain. If the stress is reapplied, the material again responds elastically at the beginning up to a new yield point *that is higher than the original yield point* (strain hardening). The amount of elastic strain that it will take before reaching the yield point is called *elastic strain recovery.*

Compressive, Shear and Torsional Deformation

Compressive and shear stresses give similar behavior to tensile stresses, but in the case of compressive stresses there is no maximum in the s-e curve, since no necking occurs.

Hardness

Hardness is the resistance to plastic deformation (e.g., a local dent or scratch). Thus, it is a measure of *plastic* deformation, as is the tensile strength, so they are well correlated. Historically, it was measured on an empirically scale, determined by the ability of a material to scratch another, diamond being the hardest and talc the softer. Now we use standard tests, where a ball or point is pressed into a material and the size of the dent is measured. There are a few different hardness tests: Rockwell, Brinell, Vickers, etc. They are popular because they are easy and non-destructive (except for the small dent).

Hardness is the resistance of a material to localized deformation. The term can apply to deformation from indentation, scratching, cutting or bending. In metals, ceramics and most polymers, the deformation considered is plastic deformation of the surface. For elastomers and some polymers, hardness is defined at the resistance to elastic deformation of the surface. The lack of a fundamental definition indicates that hardness is not be a basic property of a material, but rather a composite one with contributions from the yield strength, work hardening, true tensile strength, modulus, and others factors. Hardness measurements are widely used for the quality control of materials because they are quick and considered to be nondestructive tests when the marks or indentations produced by the test are in low stress areas.

Mohs Hardness Test

One of the oldest ways of measuring hardness was devised by the German mineralogist Friedrich Mohs in 1812. The Mohs hardness test involves observing whether a materials surface is scratched by a substance of known or defined hardness. To give numerical values to this physical property, minerals are ranked along the Mohs scale, which is composed of 10 minerals that have been given arbitrary hardness values. Mohs hardness test, while greatly facilitating the identification of minerals in the field, is not suitable for accurately gauging the hardness of industrial materials such as steel or ceramics. For engineering materials, a variety of instruments have been developed over the years to provide a precise measure of hardness. Many apply a load and measure the depth or size of the resulting indentation. Hardness can be measured on the macro-, micro- or nanoscale.

Brinell Hardness Test

The oldest of the hardness test methods in common use on engineering materials today is the Brinell hardness test. Dr. J. A. Brinell invented the Brinell test in Sweden in 1900. The Brinell test uses a desktop machine to applying a specified load to a hardened sphere of a specified diameter. The Brinell hardness number, or simply the Brinell number, is obtained by dividing the load used, in kilograms, by the measured surface area of the indentation, in square millimeters, left on the test surface. The Brinell test is frequently used to determine the hardness metal forgings and castings that have a large grain structures. The Brinell test provides a measurement over a fairly large area that is less affected by the course grain structure of these materials than are Rockwell or Vickers tests.

A wide range of materials can be tested using a Brinell test simply by varying the test load and indenter ball size. In the USA, Brinell testing is typically done on iron and steel castings using a 3000Kg test force and a 10mm diameter ball. A 1500 kilogram load is usually used for aluminum castings. Copper, brass and thin stock are frequently tested using a 500Kg test force and a 10 or 5mm ball. In Europe Brinell testing is done using a much wider range of forces and ball sizes and it is common to perform Brinell tests on small parts using a 1mm carbide ball and a test force as low as 1kg. These low load tests are commonly referred to as baby Brinell tests. The test conditions should be reported along with the Brinell hardness number. A value reported as "60 HB 10/1500/30" means that a Brinell Hardness of 60 was obtained using a 10mm diameter ball with a 1500 kilogram load applied for 30 seconds.

Rockwell Hardness Test

The Rockwell Hardness test also uses a machine to apply a specific load and then measure the depth of the resulting impression. The indenter may either be a steel ball of some specified diameter or a spherical diamond-tipped cone of 120° angle and 0.2 mm tip radius, called a brale. A minor load of 10 kg is first applied, which causes a small initial penetration to seat the indenter and remove the effects of any surface irregularities. Then, the dial is set to zero and the major load is applied. Upon removal of the major load, the depth reading is taken while the minor load is still on. The hardness number may then be read directly from the scale. The indenter and the test load used determine the hardness scale that is used (A, B, C, etc.).

For soft materials such as copper alloys, soft steel, and aluminum alloys a 1/16" diameter steel ball is used with a 100-kilogram load and the hardness is read on the "B" scale. In testing harder materials, hard cast iron and many steel alloys, a 120 degrees diamond cone is used with up to a 150 kilogram load and the hardness is read on the "C" scale. There are several Rockwell scales other than the "B" & "C" scales, (which are called the common scales). A properly reported Rockwell value will have the hardness number followed by "HR" (Hardness Rockwell) and the scale letter. For example, 50 HRB indicates that the material has a hardness reading of 50 on the B scale.

Rockwell Superficial Hardness Test

The Rockwell Superficial Hardness Tester is used to test thin materials, lightly carburized steel surfaces, or parts that might bend or crush under the conditions of the regular test. This tester uses the same indenters as the standard Rockwell tester but the loads are reduced. A minor load of 3 kilograms is used and the major load is either 15 or 45 kilograms depending on the indenter used. Using the 1/16" diameter, steel ball indenter, a "T" is added (meaning thin sheet testing) to the superficial hardness designation. An example of a superficial Rockwell hardness is 23 HR15T, which indicates the superficial hardness as 23, with a load of 15 kilograms using the steel ball.

Vickers and Knoop Microhardness Tests

The Vickers and Knoop Hardness Tests are a modification of the Brinell test and are used to measure the hardness of thin film coatings or the surface hardness of case-hardened parts. With these tests, a small diamond pyramid is pressed into the sample under loads that are much less than those used in the Brinell test. The difference between the Vickers and the Knoop Tests is simply the shape of the diamond pyramid indenter. The Vickers test uses a square pyramidal indenter which is prone to crack brittle materials. Consequently, the Knoop test using a rhombic-based (diagonal ratio 7.114:1) pyramidal indenter was developed which produces longer but shallower indentations. For the same load, Knoop indentations are about 2.8 times longer than Vickers indentations.

An applied load ranging from 10g to 1,000g is used. This low amount of load creates a small indent that must be measured under a microscope. The measurements for hard coatings like TiN must be taken at very high magnification (i.e. 1000X), because the indents are so small. The surface usually needs to be polished. The diagonals of the impression are measured, and these values are used to obtain a hardness number (VHN), usually from a lookup table or chart. The Vickers test can be used to characterize very hard materials but the hardness is measured over a very small region.

The values are expressed like 2500 HK25 (or HV25) meaning 2500 Hardness Knoop at 25 gram force load. The Knoop and Vickers hardness values differ slightly, but for hard coatings, the values are close enough to be within the measurement error and can be used interchangeably.

Scleroscope and Rebound Hardness Tests

The Scleroscope test is a very old test that involves dropping a diamond tipped hammer, which falls inside a glass tube under the force of its own weight from a fixed height, onto the test specimen. The height of the rebound travel of the hammer is measured on a graduated scale. The scale of the rebound is arbitrarily chosen and consists on Shore units, divided into 100 parts, which represent the average rebound from pure hardened high-carbon steel. The scale is continued higher than 100 to include metals having greater hardness. The Shore Scleroscope measures hardness in terms of the elasticity of the material and the hardness number depends on the height to which the hammer rebounds, the harder the material, the higher the rebound.

The Rebound Hardness Test Method is a recent advancement that builds on the Scleroscope. There are a variety of electronic instruments on the market that measure the loss of energy of the impact body. These instruments typically use a spring to accelerate a spherical, tungsten carbide tipped mass towards the surface of the test object. When the mass contacts the surface it has a specific kinetic energy and the impact produces an indentation (plastic deformation) on the surface which takes some of this energy from the impact body. The impact body will lose more energy and it rebound velocity will be less when a larger indentation is produced on softer material. The velocities of the impact body before and after impact are measured and the loss of velocity is related to Brinell, Rockwell, or other common hardness value.

Durometer Hardness Test

A Durometer is an instrument that is commonly used for measuring the indentation hardness of rubbers/elastomers and soft plastics such as polyolefin, fluoropolymer, and vinyl. A Durometer simply uses a calibrated spring to apply a specific pressure to an indenter foot. The indenter foot can be either cone or sphere shaped. An indicating device measures the depth of indentation. Durometers are available in a variety of models and the most popular testers are the Model A used for measuring softer materials and the Model D for harder materials.

Barcol Hardness Test

The Barcol hardness test obtains a hardness value by measuring the penetration of a sharp steel point under a spring load. The specimen is placed under the indenter of the Barcol hardness tester and a uniform pressure is applied until the dial indication reaches a maximum. The Barcol hardness test method is used to determine the hardness of both reinforced and non-reinforced rigid plastics and to determine the degree of cure of resins and plastics.

Chemical Properties of Materials

Some of the chemical properties of engineering materials are listed below:

1. Chemical composition
2. Atomic bonding
3. Corrosion resistance
4. Acidity or Alkalinity

Chemical Composition

The chemical composition of engineering material indicates the elements which are combined together to form that material. Chemical composition of a material affects the properties of engineering materials very much. The strength, hardness, ductility, brittleness, corrosion resistance, weldability etc. depends on chemical composition of materials.

Hence, we should also have the knowledge of chemical composition of engineering materials. For Example the Chemical compositions of some materials are listed below.

Sl. No.	Material	Chemical Composition
1.	Steel	Fe, Cr, Ni
2.	Brass	Cu = 90%, Ni = 10%
3.	Bronze	90% Cu, 10% Ni
4.	Invar	Fe = 64%, Ni = 36%
5.	Gun Metal	Cu = 88%, Tin = 10%, Zn = 2%
6.	German Silver or Nickel Silver or Electrum	Cu = 50%, Zn = 30%, Ni = 20%
7.	Nichrome	Ni = 60%, Cr = 15%, Fe = 25%
8.	Phosphor Bronge	Cu = 89 – 95.50%, Tin = 3.50 -10%, P = 1%
9.	Manganin	Cu = 84%, Mn = 12%, Ni = 4%
10.	Constantan	Cu = 60%, Ni = 40%

Atomic Bonding

Atomic bonding represents how atoms are bounded to each other to form the material. Many properties, such as melting point, boiling point, thermal conductivity and electrical conductivity of materials are governed by atomic bonding of materials. Hence, to understand the properties of materials, it is very important to study the atomic bonding of materials. Atomic bonds in materials are of following types:

- Ionic bond: Forms by exchanging of valence electrons between atoms.
- Covalent bonds: Forms by sharing of electrons between atoms.
- Metallic bonds: Found in metals.

Corrosion Resistance

Corrosion is a gradual chemical or electro-mechanical attack on a metal by its surrounding medium. Due to the corrosion, metal starts getting converted into an oxide, salt or some other compound. Corrosion of metals is effected by many factors such as air, industrial atmosphere, acid, bases, slat solutions and soils etc. Corrosion has a very adverse effect on materials. Due to corrosion, the strength and life of a material is reduced.

Corrosion resistance of a material is the ability of material to resist the oxidation in atmospheric condition. Generally pure metals such as iron, copper, aluminum etc. gets corroded in slowly in atmosphere. To avoid the corrosion of these metals in pure form, we use these metals in the form of alloys such as stainless steel, brass, bronze, German silver, Gunmetal etc.

Acidity or Alkalinity

Acidity or Alkalinity is an important chemical property of engineering materials. A material is acetic or Alkane, it is decided by the ph value of the material. Ph value of a material varies from 0 to 14. Ph value of 7 is considered to be neutral. Ordinary water is having ph value of 7. The materials which are having ph value below 7 are called Acetic and Materials which are having ph value greater than 7 are called alkane. Acidity of Alkalinity of material indicates that how they react with other materials.

Electrical Properties of Materials

The Electrical properties of a material are those which determine ability of material to be suitable for a particular Electrical Engineering Application. Some of the typical Electrical properties of engineering materials are listed below:

- Resistivity
- Conductivity
- Temperature coefficient of Resistance
- Permittivity
- Thermoelectricity

Resistivity

It the property of material which resists the flow of electric current through material. It is the reciprocal of conductivity.

It is dented by ‘ρ’. Resistivity of a material of a conductor can be determined as below:

$$\rho = R\frac{A}{l}$$

where,

- ‘R’ is the resistance of conductor in Ω.
- ‘A’ is the cross sectional area of conductor in m^2.
- ‘l’ is the length of the conductor in meter SI unit of resistivity of is Ω¦-meter. Resistivity of some materials is listed below.

Sl. No.	Element	Resistivity at 20°C in Ω – m
1	Silver	1.59×10^{-8}
2	Copper	1.7×10^{-8}
3	Gold	2.44×10^{-8}
4	Aluminum	2.82×10^{-8}
5	Tungsten	5.6×10^{-8}
6	Iron	1.0×10^{-7}
7	Platinum	1.1×10^{-7}
8	Lead	2.2×10^{-7}
9	Manganin	4.82×10^{-7}
10	Constantan	4.9×10^{-7}
11	Mercury	9.8×10^{-7}
12	Carbon (Graphite)	3.5×10^{-5}
13	Germanium	4.6×10^{-1}
14	Silicon	6.4×10^{2}
15	Glass	10^{10} to 10^{14}
16	Quartz (fused)	7.5×10^{17}

Conductivity

It is the property of material with allows the flow of electric current through material. It is a parameter which indicates that how easily electric current can flow through the material. It is denoted by 'σ'. Conductivity of material is the reciprocal of resistivity. Conductivity of material can be determined by,

$$\sigma = \frac{1}{\rho} = \frac{I}{R.A}$$

Its SI unit is 1/(Ω-meter) or ℧/meter.

Dielectric Strength

It is the property of material which indicates the ability of material to withstand at high voltages. Generally it is specified for insulating material to represent their operating voltage. A material having high dielectric strength can withstand at high voltages. Generally, it is represented in the unit of KV/cm. Dielectric strength of some insulating materials are listed below.

Sl. No.	Material	Dielectric Strength [KV(max.)/cm]
1	Air	30
2	Porcelain	80
3	Paraffin Wax	120
4	Transformer oil	160
5	Bakelite	220
6	Rubber	280
7	Paper	500
8	Teflon	600
9	Glass	1200
10	Mica	2000

Temperature Coefficient of Resistance

The temperature coefficient of resistance of a material indicates the change in resistance of material with change in temperature. Resistance of conductor changes with change of temperature.

The rise in resistance of a material with rise in temperature depends on following things:

1. $R_2 - R_1 \propto R_1$,
2. $R_2 - R_1 \propto t_2 - t_1$,
3. Property of material of conductor.

Where, R_1 is the resistance of conductor at temperature of t_1 °C and R_2 is the resistance of conductor at temperature of t_2 °C.

Hence, from above,

$$R_2 - R_1 \propto R_1 (t_2 - t_1)$$

Or,

$$R_2 - R_1 = \alpha_1 R_1 (t_2 - t_1) \Rightarrow R_2 = R_1 [1 + \alpha_1 (t_2 - t_1)]$$

Where, α_1 is temperature coefficient of resistance of material at temperature of t_1°C. Its unit is / °C. Temperature coefficient of resistance of material is also depends on temperature. Temperature coefficient of some materials are listed below.

Sl. No.	Element	Temperature Coefficient of Resistance in /°C
1	Manganin	0.00002
2	Constantan	0.00017
3	Nichrome	0.0004
4	Mercury	0.0009
5	Silver	0.0038
6	Copper	0.00386
7	Annealed copper	0.000393
8	Platinum	0.003927
9	Aluminum	0.00429
10	Carbon (Graphite)	– 0.0005
11	Germanium	– 0.05
12	Silicon	– 0.07

Thermoelectricity

If the junction, formed by joining to two metals, is heated, a small voltage in the range of millivolt is produced. This effect is called thermoelectricity or thermoelectric effect. This effect forms the basis of operation of thermocouples and some temperature based transducers. This effect can be used to generate electricity, to measure the temperature and to measure the change is temperature of objects.

Thermal Properties of Materials

Specific Heat (Heat Capacity)

The heat capacity of a material is defined as the amount of heat required to raise its temperature by 1°. The heat capacity per unit mass, of material is defined as its specific heat. Heat capacity per mole is defined as its molar heat capacity.

Mathematically, specific heat of a solid is defined as:

$$c = \frac{1}{m}, \frac{dQ}{dT}$$

Where,

- m = Mass,
- T = Temperature,
- Q = Energy content, and
- dQ = Energy (heat) added or subtracted to produce the temperature change dT.

For unit mass per degree change in temperature specific heat c = dQ, the quantity of heat that must be added per unit mass of a solid to raise its temperature by one degree. The specific heat of material is sometimes defined as the ratio of its heat capacity to that of water. Specific heat in this becomes the dimensionless unit (as specific heat of water is unity in MKS units).

For gases there are two specific heats i.e., specific heat at constant volume c_v and specific at constant pressure c_p. c_p is always greater than c_v since any substance expands on heating and extra heat is required to raise the temperature by 1 degree in order to compensate for the energy required for expansion. For solids, difference between c_p and c_v is negligible and only one specific heat is used ($c_p = c_v = c$). This is due to the fact that in solids and liquids the expansion with heating is very small.

According to classical kinetic theory of heat, heat capacity of an atom in a solid (crystalline element) is constant and is equal to 26 kJ/kg atoms (°C) at room temperature. This is to be divided by molecular weight in order to get mass specific heat of a solid.

Representative specific heat at room temperature (20°) are given below (units)(kJ/ kg/° C):					
Metals		Ceramic		Polymers	
Nickel	0.452	Gas	0.711	Polyethylene	2.3
Iron	0.447	Alumina	0.84	Nylon	1.67
Lead	0.31	Zirconiumoxide	0.46	Acrylics	1.46
Zinc	0.393				
Copper	0.385				
Tin	0.226				
Aluminium	0.9				
Magnesium	1.04				
Lithium	3.3				
Uranium	0.117				

Specific heat increases slightly with increase in temperature and varies from metal to metal. An increase of 5 percent for every 100 °C temperature rise can be used as a general approximation.

The effect of raising temperature of metals and alloys is to raise the amplitude of vibration of each atom and the heat energy so absorbed is the specific heat.

Thermal Conductivity

It is defined as the amount of heat conducted in a unit time through a unit area normal to the direction of heat flow. Heat conduction through isotropic solids is expressed by Fourier's law:

$$q = -k\frac{dT}{dx}$$

where,

- q = Rate of heat flow/unit area normal to the direction of flow,
- T = Temperature,
- x = Distance measured in the direction of flow, and
- k = Thermal conductivity.

Heat flow through solids is due to elastic vibration of atoms or molecules or due to transfer of energy by the free electrons. Metals have large supply of free electrons which account for their thermal conductivity. Both types of conduction occur in metals and semiconductors. Insulators have lower conductivities as they depend entirely on the lattice vibration of atoms and molecules. This is a slower process than electronic conduction.

The theory of thermal conductivity through crystalline solids (metals) based on quantum (solid state) theory can be explained by concept of phonons which represent the particles (gas) characteristics of a thermal wave. It is a quantum of energy and vibration of a thermoelastic (acoustic) wave.

In dielectrics (thermal insulators) thermal conductivity is caused alone by the atomic or molecular vibration of the lattice (lattice is a geometrical array of lines or points in which atoms are considered spheres) representing a certain type of crystal (say metal) structure.

The progress of this elastic thermal wave (or phonons) through a crystal is akin to a gas molecule through a gas. At a heated surface the motion is increased so that collision with other phonons occurs at an increased rate and thus heat is transmitted to other parts of the phonon gas. Thermal conductivity in solids is given by a formula similar to that derived from the kinetic theory of gases:

$$k = \frac{1}{3}cv\lambda$$

where,

- k = Thermal conductivity,
- c = Specific heat per unit volume,

- v = Average particles velocity or velocity of the lattice wave (the velocity of sound), and
- λ = Mean free path of lattice wave (phonon) of a given frequency.

In an ideal crystal, the atomic or molecular waves of vibration are harmonic, hence, X is very large and it should have infinite thermal conductivity. In actual crystals mutual scattering and lattice wave (phonons) may occur, due to inharmonicity of the vibration and internal crystal imperfection. Phonons scattering and thus thermal conductivity depends, on crystalline structure of metals and alloys.

A comparison of thermal and electrical conductivities is given below:

Metals	Thermal conductivity	Electrical conductivity
Copper	100%	100%
Nickel	23%	25%
Zinc	29%	29%
Tin	17%	15%
Lead	9%	8%

Some typical thermal conductivities are shown as follows:

Material	Conductivity (cal / sec) cm
Silver	1.001
Alminium	0.485
Iron	0.174
Antinomy	0.004
Cadmium	0.223
Copper	0.927
Lead	0.084
Magnesium	0.270
Molybdenum	0.349
α Nickel	0.157
Platinum	0.166
Tin	0.157
Tungsten	0.382
Zinc	0.270
Titanium Alloy	0.021
Glasses	0.0025
Silicon Martics	0.0022
Acrylics	0.0005
Foamed polystyrene	0.001

The thermal conductivity of pure metals increases as temperature is lowered often to a considerable degree. Copper has thermal conductivity about 35 times greater at – 269 °C than at 20 °C.

Alloys, however, do not show this pronounced increase of thermal conductivity at lower temperatures and only small percentages of alloying are required to suppress this change in thermal characteristics.

At normal and elevated temperatures, pure metals and their alloys possess very low temperature co-efficient of thermal conductivity and thus for all design purposes these effects of higher temperature on thermal conductivity are usually ignored.

The thermal conductivity of amorphous solids such as glasses, and plastics increases with a rise in temperature. They generally possess, low thermal conductivity at room temperature. This is due to the fact that amorphous solids have excessive. scattering of phonons by their disordered structure at lower temperatures.

The thermal conductivity of refractories (more complex solids) depends on their chemical composition and crystalline structures. This is due to the presence of impurities and comparatively smaller grain size and porosity which result in lower values of thermal conductivity.

If structure is simple as in case of silicon carbide, thermal conductivity has higher value. Fire clay bricks and fuel fused silica also show an increase in thermal conductivity with increasing temperature. On the other hand in case of magnesite and alumina which are more crystalline in nature, the thermal conductivity decreases with rising temperature.

Thermal Expansion

Thermal expansion arises from the addition of heat energy in the atoms and their subsequent movement away from their equilibrium positions as the temperature rises in solid. This expansion or contraction resulting from increase or decrease in temperature is three dimensional but in practice linear thermal expansion is used for simplicity instead of volume expansion.

The increase in length per unit length per degree rise in temperature is called coefficient of linear expansion. Thermal expansion does not necessarily vary uniformly with temperature but it is sufficiently linear over narrow ranges of temperature.

If the bonds between the atoms are strong and highly directional as in ionic and covalent solids, the thermal expansion will be relatively small. If on the other hand the atoms are more loosely bound as in metals, a greater degree of expansion is there. In molecular solid, where bonding least resists the movement of the molecules, the thermal expansion will be the greatest.

The thermal expansion of solid is related to other thermal properties such as specific heat and melting point as all these properties have their origin in lattice vibrations which increase with the temperature. The atoms or molecules as earlier explained oscillate (vibrate) with a certain amplitude about their equilibrium positions.

The amplitude of this vibration increases as the temperature rises resulting in moving further away of atoms and molecules from their equilibrium position causing an increase in volume (or linear expansion) of solid. In this way magnitude of the coefficient of thermal expansion of solids will depend on their interatomic and intermolecular forms and also on their structural arrangement.

It has been observed that between absolute zero temperature and the melting point, total volume range of elements is approximately constant. This can be interpreted that materials with lower softening (melting) points will have higher expansion coefficients. This also means that thermal expansion will approach zero at the absolute zero temperature.

Organic polymers such as plastics and rubber have many times higher expansion coefficients than metals because of their relatively lower softening point. This may be reduced by addition of filler materials (such as glass fibre, asbestos, alumina etc.) possessing lower thermal expansion coefficients. Alloying of metals has a minor effect on this property.

Melting Point

Melting point or softening point is a significant temperature level as it represents transition point between solid and liquid phases having different structural arrangement of the atoms within the material. As heat is added to a solid, its thermal energy increases until the atoms or molecules on the surface begin to break away from their equilibrium positions.

There is a link between interatomic spacing at which the bonding force is maximum and the amplitude of thermal vibration at which this breaking away occurs as if the atoms can be separated at this point, no further increase in force is needed to separate them further. After melting commences, any further heat is all used up in activating more particles of solids which in turn collide with neighbouring particles transmitting their energy to them.

The structure is therefore transformed from a solid having definite equilibrium positions to a liquid having only short range order. During melting no further rise in temperature occurs and solid and liquid phases exist at the same temperature. Melting temperature depends upon the amount of thermal energy required.

This in turn depends on the nature of interatomic and intermolecular bonds. Therefore higher melting point is exhibited by those materials possessing stronger bonds. Covalent, ionic, metallic and molecular types of solids have decreasing order of bonding strength and thus the melting points.

Crystalline solids have a sharp melting point at which there is sudden transformation from solids to liquid states. Amorphous solids such as glasses, plastics and rubbers and also clays do not have definite melting points but soften gradually over a certain temperature range.

Relation between Thermal Expansion and Melting Point

Both depend upon the bonds between atoms (or molecules) of the solid and so are related. For each class of materials,

$$\alpha T_m = \text{constant}$$

where,

- α = Coefficient of thermal expansion, and
- T_m = Melting temperature.

Therefore, any two materials of a given class possessing same coefficient of expansion will therefore have approximately same melting point.

The value of this constant is as under:

Class of material	Approx. value of $\propto T_m$
(i) Metal and ionic compound and certain salts :	0.02 to 0.03
(ii) Covalent bonded oxides and glasses	0.007

There is an interesting conclusion that for a material to be coated to another material, coating will have to be of different class than the base material if both must have same thermal expansion.

Heat Resistance

Melting point determines the heat resistance of a material as any material for high temperature application should have its melting point above the service temperature. Ceramic materials are known to have high melting points and good chemical stability but they are difficult to fabricate and cannot take thermal or mechanical shock.

Following is the list of some materials possessing resistance to high temperatures:

Metallic oxides	Metallic carbides	Miscellaneous
Alumina	Nitrides	Graphites
Silica	Borides	Molybdenum
Zirconia	Silicides	Titanium
Magnesia	A few sulphiades	Tungsten Zirconium

Thermal Shock

Thermal shock is the effect of a sudden change of temperature on a material whereas thermal shock resistance can be defined as the ability of material to withstand thermal stresses due to sudden and severe changes in the temperature at the surface of a solid body.

If a solid structure is prevented so that it cannot expand or contract freely on heating or cooling, excessive thermal stresses may result culminating in thermal shock and causing failure of the body. Thermal shock resulting from cooling which results in tensile stresses at the surface is much more dangerous than that from heating.

Thermal shock resistance of a solid is sometimes given by the equation:

$$\text{Thermal shock resistance} = k\frac{\sigma_t}{E\alpha}$$

where,

- k = Thermal conductivity,
- σ_t = Tensile strength,
- E = Young's modulus, and
- α = Linear co-efficient of thermal expansion.

For maximum shock resistance:

- Thermal-conductivity should be high.
- Thermal expansion should be low.
- Material should have low elastic modulus and high tensile strength.

Brittle materials such as glass and ceramics are particularly prone to thermal shock because they readily experience brittle failure instead of plastic yield.

Thermal Diffusivity

Thermal diffusivity (h) is defined as:

$$h(\text{ Thermal diffusivity}) = \frac{k(\text{conductivity})}{c_p(\text{heat capacity})\,\rho(\text{density})}$$

$c_p\ \rho$ represent heat requirement per unit volume. A material having high heat requirement per unit volume possesses a low thermal diffusivity because more heat must be added to or removed from the material for affecting a temperature change. Thermal diffusivity is therefore associated with the diffusion of thermal energy and may be taken to represent an energy flux arising from the motion of phonons through a relatively stationary atomic array. As phonons are in the nature of waveform, the atoms vibrate in unison but are not physically transported.

Thermal Stresses

When expansion or contraction of a body due to temperature change is wholly or partially prevented, thermal stress will be developed in body. Thermal stress may arise from external bodies connected to one under stress as for example, welded structure, railway line shrink fit components. Or, it may be due to non-uniform expansion of the body itself, for example bimetallic strips used in thermostatic controls. The value of thermal stress, expansion or contraction can be calculated by applying simple stress calculation theory.

Thermo-Elastic Effect

When a solid is subjected to a load, work is done on it and it changes in volume. If this work is done at constant temperature, an adiabatic temperature rise (without transfer of heat to or from the surroundings) occurs. This will appear in the form of rise of temperature of solid when it is in stretched condition. Similarly when the solid is rapidly relaxed, it will feel cool. This warming or cooling phenomenon is called thermoelastic effect.

Optical Properties of Materials

Reflection and Transmission of Light

Normal to the Interface

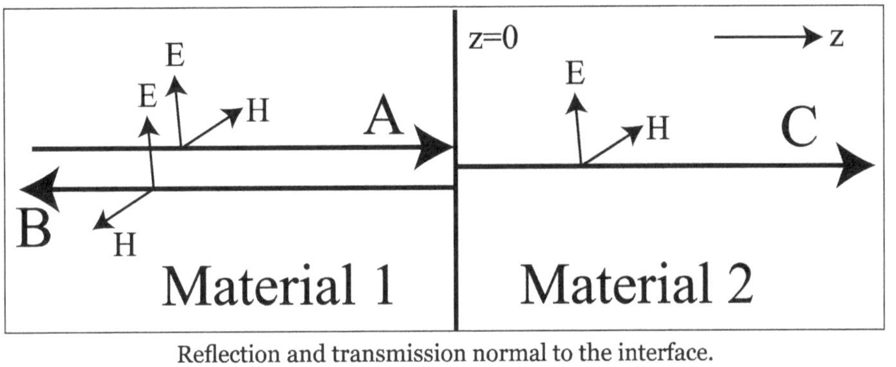

Reflection and transmission normal to the interface.

We shall now examine the properties of reflected and transmitted electromagnetic waves. Let's begin by considering the case of reflection and transmission where the incident wave is normal to the interface, as shown in figure. We have been using $E = E_0 \exp[i(kz - \omega t)]$ to describe waves, where the speed of the wave (a function of the material) is ω/k. In order to express the wave in terms which are explicitly a function of the material in which it is propagating, we can write the wavenumber k as:

$$k = \frac{\omega}{c} = \frac{\omega_n}{c_0} = nk_0$$

where, n is the refractive index of the material and k_o is the wavenumber of the wave, were it to be travelling in free space. We will therefore write a wave as the following (with z replaced with whatever direction the wave is travelling in):

$$Ex = E_0 \exp\left[i\left(nk_0 z - \omega t\right)\right]$$

In this problem there are three waves we must consider: the incident wave A, the reflected wave B and the transmitted wave C. We'll write down the equations for each of these waves, using the impedance $Z = E_x/H_y$, and noting the flip in the direction of H upon reflection, and the different refractive indices and impedances for the different materials.

$$E(A) = E_A e^{i(n_1 k_0 z - \omega t)}, \quad E(B) = E_B e^{i(n_1 k_0 z - \omega t)}, \quad E(C) = E_C e^{i(n_2 k_0 z - \omega t)}$$

$$H(A) = \frac{E_A}{Z_1} e^{i(n_1 k_{\rightarrow} z - \omega t)}, \quad H(B) = -\frac{E_B}{Z_1} e^{i(n_1 k_0 z - \omega t)}, \quad H(C) = \frac{E_C}{Z_2} e^{i(n_2 k_0 z - \omega t)}$$

Optical Properties of Materials

At the boundary (z = 0), and for any time t, the total E field on each side of the interface must be equal (and similarly for H). Thus:

$$E_B = E_C \ and \ \frac{E_A}{Z_1} - \frac{E_B}{Z_1} = E_C$$

Rearranging, we obtain the ratio of the reflected and transmitted amplitudes:

Reflected: $$\frac{E_B}{E_A} = \frac{Z_2 - Z_1}{Z_2 + Z_1}$$

Transmitted: $$\frac{E_c}{E_A} = \frac{2Z_2}{Z_2 + Z_1}$$

For the case where the relative permeability μ_r = 1, a valid assumption for most materials at optical frequencies, we can express1 this in terms of the refractive index n.

Reflected: $$\frac{E_B}{E_A} = \frac{Z_2 - Z_1}{Z_2 + Z_1} = \frac{n_1 - n_2}{n_1 - n_2}$$

Transmitted: $$\frac{E_c}{E_A} = \frac{2Z_2}{Z_2 + Z_1} = \frac{2n_1}{n_1 - n_2}$$

Note that these represent the field amplitudes, and not power (which goes as field squared). Similar expressions can be derived for the ratio of the magnetizing fields H, using the relation that H $A,B = EA,B/Z_1$ and $H_C = E_c/Z_2$. These expressions should look reminiscent of those derived in your quantum mechanics course for a particle incident on a potential barrier.

At an Angle θ to the Interface

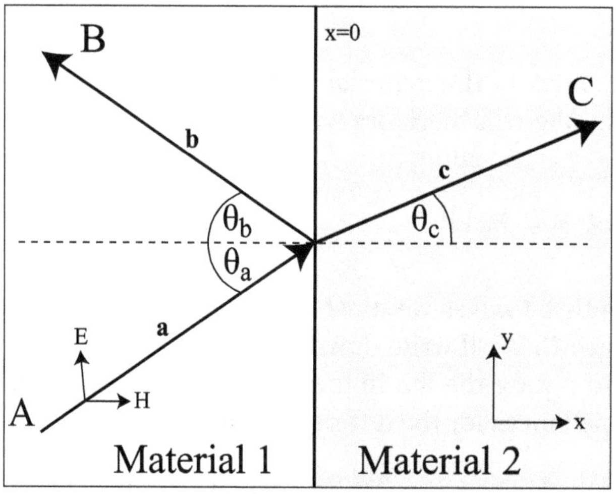

Reflection and transmission at an interface.

In figure shows a diagram of an electromagnetic wave incident on some interface between two materials 1 and 2. Note that for this derivation we must break with our convention of the waves travelling along z. So, for this part, the wave will be travelling in the $x - y$ plane. z is perpendicular and coming out of the page. Here, the three waves *A, B* and *C* travel along directions $\vec{a}, \vec{b}$ and $\vec{c}$ respectively:

$$E(A) = E_A exp\left[i\left(n_1 k_0 \vec{a} - \omega t\right)\right]$$

$$E(B) = E_B exp\left[i\left(n_1 k_0 \vec{b} - \omega t\right)\right]$$

$$E(C) = E_C exp\left[i\left(n_2 k_0 \vec{c} - \omega t\right)\right]$$

$$\vec{a} = \cos\,\theta_a \vec{x} + \sin\,\theta_a \vec{y}$$

$$\vec{b} = -\cos\,\theta_b \vec{x} + \sin\,\theta_b \vec{y}$$

$$\vec{c} = \cos\,\theta_a \vec{x} + \sin\,\theta_c \vec{y}$$

There is the boundary condition that along the interface ($x = o$), all three must remain in phase. Hence, their arguments must be equal:

$$i\left(n_1 k_0 \sin\theta_a \vec{y} - wt\right) = i\left(n_1 k_0 \sin\theta_b \vec{y} - wt\right)$$
$$= i\left(n_2 k_0 \sin\theta_c \vec{y} - wt\right)$$

Equating the arguments of these terms, we see that the frequency ω must remain constant, and that for the reflected wave:

$$n\theta_a = n_1 sin\ \theta_b \text{ and } s_o\, \theta_b = \theta_a,$$

which is known as the Law of Reflection. While for the transmitted wave:

$$\theta_2 = n_2\ \sin\theta_c \text{ and so } \sin\theta_c = n_1\ n_2\ \sin\ \theta_a$$

For $\frac{n_1}{n_2}\sin\theta_a \leq 1$, this reduces to $\theta_c = \frac{n_1}{n_2}\theta_a$ which is known as Snell's Law (for refraction) However, for $\frac{n_1}{n_2}\sin\theta_a > 1$, there is no allowed value of θ_c, and there is no transmission (this shall be useful in total internal reflection in optical fibres later on).

For the amplitudes, the algebra gets a little bit more cumbersome, but this is derived in detail in Bleaney & Bleaney. Here we will simply quote the results, for the case where μr = 1.

$$-\frac{E_B(x)}{E_A(x)} = \frac{E_B(y)}{E_A(y)} = \frac{\sin 2\theta_c - \sin 2\theta}{\sin 2\theta_c + \sin 2\theta} \quad \text{reflected}$$

$$\frac{E_C(x)}{E_A(x)} = \frac{E_C(y)}{E_A(y)} = \frac{4 \sin \theta_c \cos \theta}{\sin 2\theta_c + \sin 2\theta} \quad \text{transmitted.}$$

Note that on reflection, the *x*- and *y*-components of the electric field *E* go to zero for the case where $\tan \theta = n_2 / n_1$, which is known as the Brewster angle θ_B. The component of *E* along *z* does not go to 0, and so an electromagnetic wave becomes plane polarised when incident at the Brewster angle.

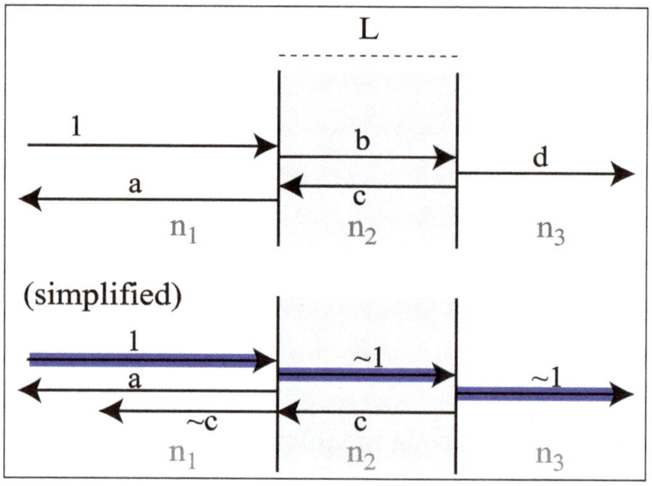

Coating an interface to reduce reflection. The lower diagram shows a simplified way of approaching the problem where reflection is assumed to be small.

Lens Blooming

The fact that there must be some reflection at any interface means that we must use interference between two or more beams to reduce it to zero. The general case is indicated schematically in figure for a coating or bloom of refractive index n_2 on the lens of index n_3. Using the results above, we note the amplitude reflection and transmission coefficients are:

$$= \frac{n_i - n_j}{n_i + n_j} \text{ and } t_{ij} = \frac{2_{ni}}{n_i + n_j}$$

We can simplify the problem by assuming reflection is already reasonably small. For zero reflection we need to ensure that the two components marked *a* and *c* are equal in amplitude and π out of phase so as to destructively interfere. The amplitude of $a = r_{12}$ and that of $c = r_{23}$, so these reflections coefficients will have to be equal for any chance of complete destructive interference. Thus:

$$= \frac{n_i - n_j}{n_i + n_j} \text{ and } t_{ij} = \frac{2_{ni}}{n_i + n_j}$$

This gives the solution $n_2 = \sqrt{n_1 n_3}$, so we know what the refractive index n_2 of the coating must be. But what about the thickness? Wave *c* has travelled an additional length *2L*, and so will have picked up an extra phase (think about the exp $(i n_i k_o z)$ term for a wave):

$$\phi = 2Ln_2k_0$$

For complete cancellation, we need this phase difference to be π:

$$2Ln_2k_0 = \pi$$

$$L = \frac{1\pi}{n_2\ 2k_0} = \frac{1}{n_2}\frac{\lambda_0}{4} = \frac{\lambda_2}{4},$$

where, $\lambda_0 = 2\pi/k_0$ is the wavelength of the wave, were it to be travelling in free space and λ_2 is the wavelength of the wave when in material 2. Thus, blooming with a layer of the appropriate thickness will remove all reflection at a particular wavelength. This is also called a 'quarter-wave' coating.

Dielectric Mirrors

What if we want to do the opposite, and have only reflection — i.e. create a mirror? This can be achieved using multiple layers of very low loss dielectric, as shown in figure. The total reflected wave is the sum of reflections from each interface in the layered structure. When the refractive index of the incident material is less than the current one ($n_j > n_{j+1}$), the reflection coefficient $r_{j,j+1}$ is positive and there no phase shift on reflection. In the opposite case ($n_j < n_{j+1}$), the reflection coefficient $r_{j,j+1}$ is negative and reflection is accompanied by an instantaneous π phase shift in the wave. Let's count up all the contributions to the reflected wave (assuming $n_1 > n_2 > n_0$):

	Phase		Reflection		Propagation
Reflection at entrance	ϕ_1	=	π	+	0
From interface 2	ϕ_2	=	0	+	$2L_1n_1k_0$
From interface 3	ϕ_3	=	π	+	$\phi_2 + 2L_2n_2k_0$
From interface 4	ϕ_4	=	0	+	$\phi_2 + 2(L_1n_1 + L_2n_2)k_0$
.					
.					
From interface 2n+1	ϕ_{2n+1}	=	π	+	$\phi_{2n} + 2L_2n_2k_0$
From interface 2n+2	ϕ_{2n+1}	=	0	+	$\phi_{2n} + 2(L_1n_1 + L_2n_2)k_0$

If we set both layers to 'quarter-wave' thickness (i.e. $L_1 = \lambda_1/4$ and $L_2 = \lambda_2/4$), then:

$$\phi_1 = \phi_2,\quad \phi_{2n+1} = \phi_{2n} + 2\pi \quad \text{and} \quad \phi 2_{n+2} = \varphi_{2n} + 2\pi$$

In other words, all waves are in phase and the assembly acts as a mirror. Efficiencies can be very high (e.g. 99.8% compared with 98.5% for the best silvered mirror).

If instead we used 'half-wave' layers ($L_i = \lambda_i / 2$), then we'd have:

$$\phi_2 = \phi_1 + \pi,\quad \phi_{2n+1} = \phi_{2n} + 3\pi \text{ and } \phi_{2n+2} = \phi_{2n} + 4\pi$$

The waves are alternately in anti-phase (phase difference π) and the system is non-reflecting.

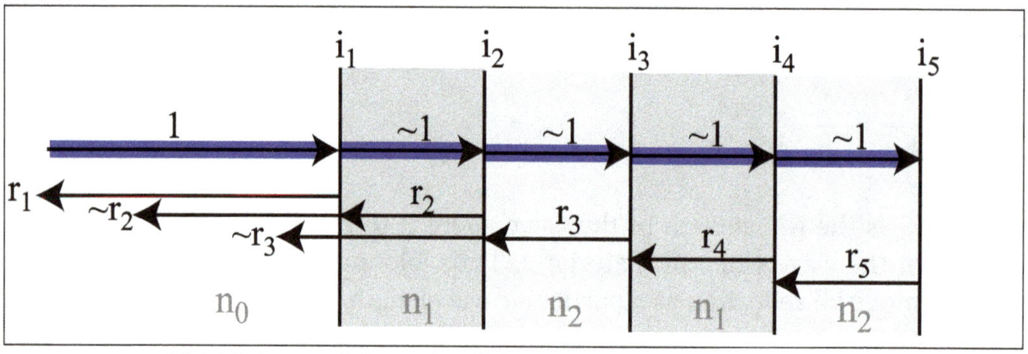

Multiple layers of alternating dielectric to create a dielectric mirror.

Optic Fibres

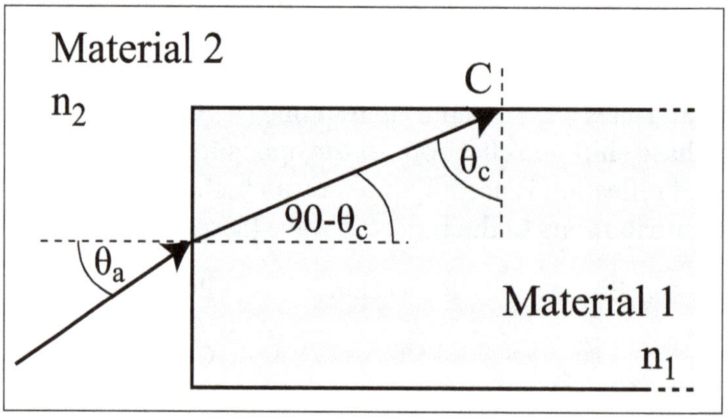

Basic theory of light pipes.

Optic fibres or 'light pipes' operate on the principle of total internal reflection. Above we showed that there is no transmission at an interface (from region 1 into region 2) when the angle of incidence θ is such that:

$$\frac{n_1}{n_2}\sin\theta > 1$$

Take some light entering a light pipe with an angle of incidence θ_a, as shown in figure At its front surface, light is accepted into the material provided:

$$\sin\theta_a \leq \frac{n_1}{n_2}, \text{ with } \sin(90 - \theta_c) = \cos\theta_c = \frac{n_2}{n_1}\sin\theta_a$$

The light remains trapped within the material if:

$$\sin\theta_c > \frac{n_2}{n_1}$$

Note that this inequality is satisfied for all θ if $\frac{n_1}{n_2} > \sqrt{2}$, such that all incident light would be accepted into the guided cone.

Provided that angles of incidence are well within these limits, the light may be guided around curved paths, such as illustrated in figure.

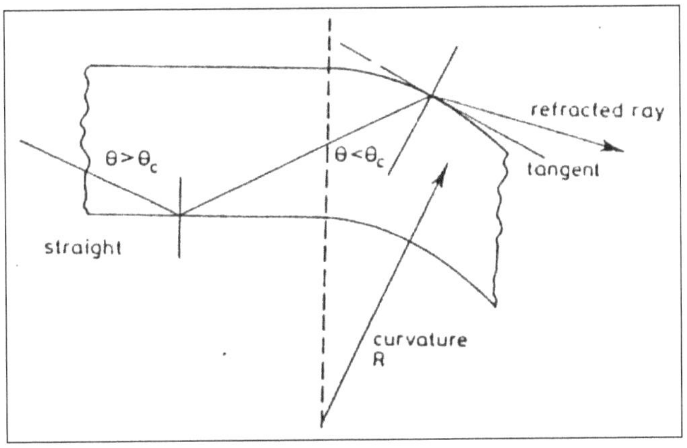

A curved light pipe.

For most real optic fibers, the surrounding material (with n_2) is not air, but rather some cladding to provide mechanical support to the light pipe. The basic requirements for an optical fibre are:

- High core refractive index (n_1).
- Low cladding refractive index (n_2).
- Low core dispersion.
- Smooth interface between core and cladding.
- Sufficient cladding thickness.
- Good flexibility.
- Good transmission efficiency.

The requirement for low dispersion (the constancy of refractive index) becomes important when the light pipe is used for telecommunications etc., when modulated signals of slightly different frequencies must not spread out as they travel along the pipe. The smooth interface is to prevent leakage and scattering from imperfections. The requirement of sufficient cladding thickness is to prevent leakage from the evanescent wave present in the cladding. A thick cladding layer is also handy for protection. Good flexibility is an obvious requirement for something to be mechanically stable, while good transmission efficiency is essentially saying the imaginary part of the refractive index must be small so that there is negligible decay as the light passes through the material. For $\mu_r \sim 1$,

$$n = \sqrt{}r\ \mu r \sim \sqrt{}$$

so this is equivalent to saying the relative permittivity $\in_r$ must be real, positive. Low absorption tends to go with low dispersion, so a good dielectric benefits both criteria.

In practice, standard glass and silica fibres are adequate for optical frequencies and the short distances involved in medical applications, (Christmas trees) etc., but significant improvements have been necessary for the multi-km distances needed by telecommunications. These improvements have been required in the whole system, from the emitter to connection to fibre to connection to detector.

The attenuation (or absorption) characteristics of a typical fibre as a function of wavelength are shown in figure. Also shown in the figure is which source and detector materials are available for different wavelengths. Note the windows of good transmission in the near infrared at 1.2 and 1.55 μm, accessible by use of an InGaAs ternary and InGaAsP quaternary semiconductors.

A typical emitter/fibre system is shown in figure, with dimensions indicated.

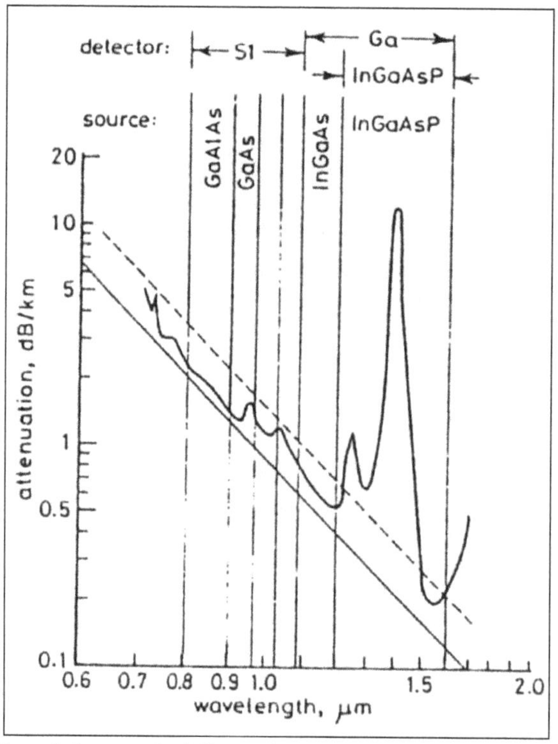

Attenuation versus wavelength for a typical fibre. Also shown are some available sources and detectors.

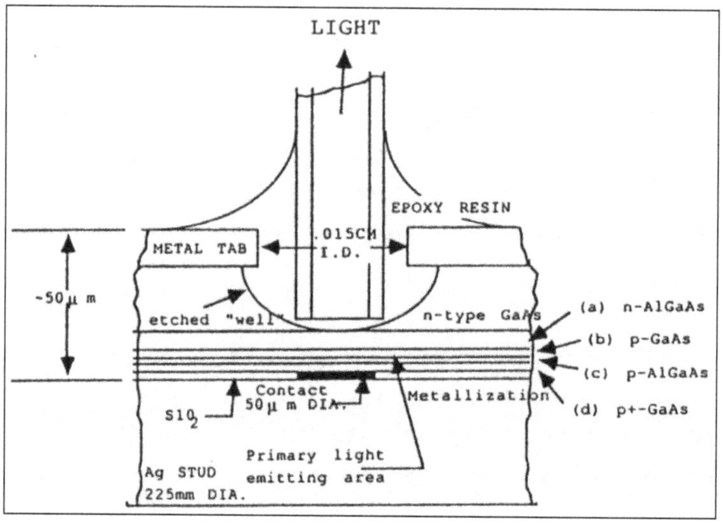

A typical LED emitter fibre system.

Polarised Light

We have assumed the electric and magnetic field are fixed along x and y, respectively, perpendicular to the direction of propagation z. In this sense we have already been assuming one particular polarisation (though everything we've done could be generalised). The kind of polarisation we have used is plane polarised, and we could rotate E_x and H_y around the $x - y$ plane to generate other kinds of plane polarisation.

Ordinary (unpolarised) light has equal intensity of electric field vectors along both x and y and so there is no net polarisation, and can be thought of as a combination of two light beams of orthogonal polarisations.

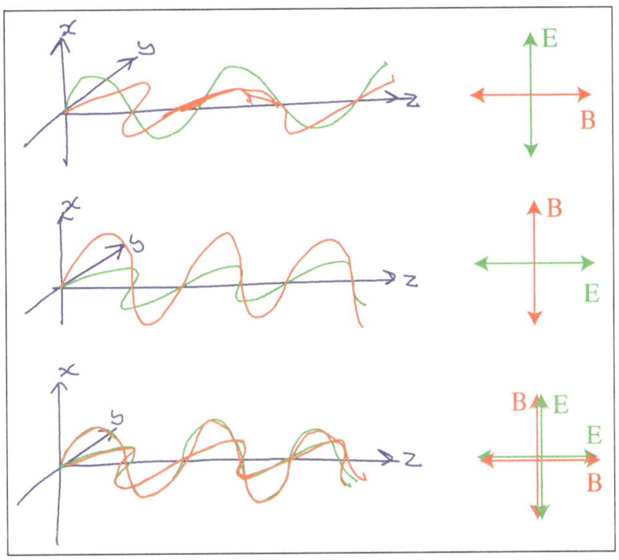

Pictorial representation of plane polarised and ordinary light beams.

Selective Reflection/Transmission

We found when treating the problem of light incident at some angle θ to an interface, that there was a particular angle, the Brewtser angle: $\tan \Phi_B = n_2/n_1$, where only the component of the electric field perpendicular to the directions of propagation remains. Therefore, incident light at the Brewster angle becomes plane-polarised in reflection. For a typical glass with refractive index 1.54, the Brewster angle is around 57°, as shown in figure.

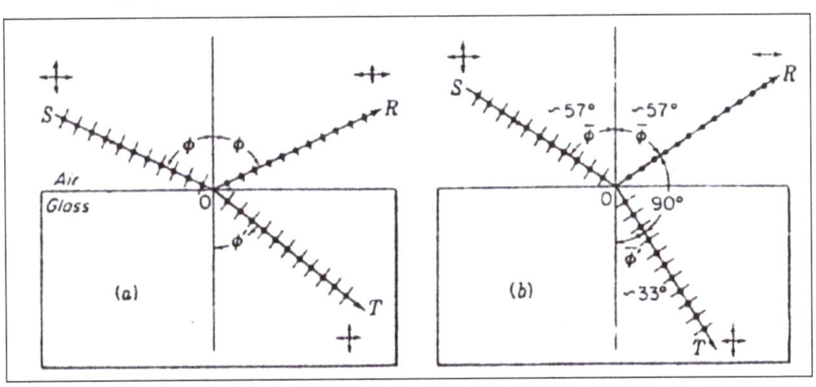

(a) Partial polarisation of light by reflection and transmission.
(b) Complete polarisation in reflection at the Brewster angle.

The reflected wave, although polarised, is of rather small amplitude. The transmitted wave is partially polarised and, to a first approximation, retains the full amplitude of the incident light (at that polarisation). Thus, as an alternative, polarisation may be obtained as an increasing fraction of the transmitted wave, as illustrated in figure.

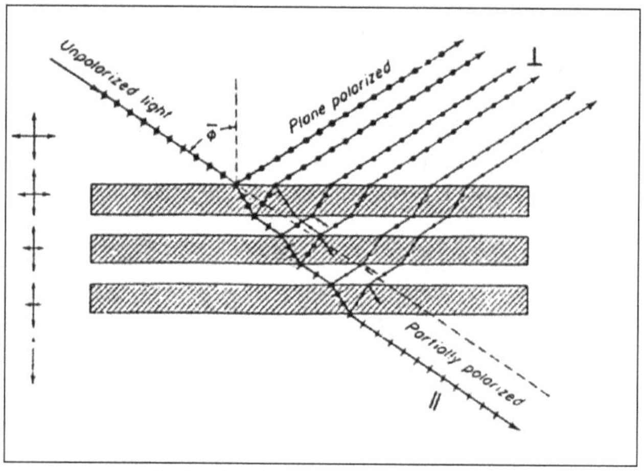

Polarisation by a pile of glass plates.

Double Diffraction

Birefringence is the property of a material whereby the refractive index depends on the polarisation of the light. This must rely on some non-cubic symmetry in the crystal structure (e.g. tetragonal, hexagonal or trigonal, with the two directions of interest being parallel or perpendicular to the c-axis). When unpolarised light enters such a crystal, it splits into two plane-polarised portions which then propagate according to the two different refractive indices. This opens up several possibilities for their separation, illustrated by the Nicol prism in figure. The incoming light *S* splits into the ordinary *O* and extraordinary *E* rays inside the calcite crystal, with refractive indices $n(O) = 1.66$ and $n(E) = 1.49$. Two pieces of the crystal are glued together using Canada balsam, which is not birefringent and has $n = 1.55$). Thus, at an appropriate range of angles, one of the rays (*O* say) is totally internally reflected while the other (*E* say) is transmitted. Note that the reflected ray contains *O* and a component of *E* which was reflected also at the interface, so it contains only partial polarisation.

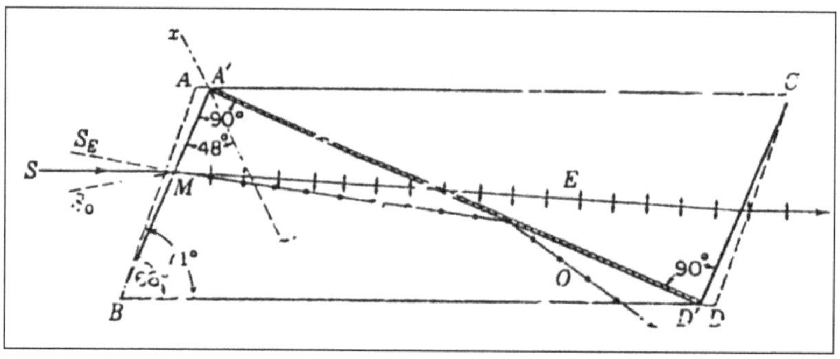

The Nicol prism for the production of polarised light.

Selective Absorption

The phenomenon of dichroism refers to a material absorbing light at different strengths depending on its polarisation. A commercial implementation of this came from Polaroid who developed the system of dichroic crystals in a plastic sheet which were aligned during production. A pair of parallel Polaroid sheets transmits about 40% of the incident light (of the possible 50%) and less than 0.01% when crossed (i.e. at 90°).

Transmission through a Polariser and Analyzer

We may take two polarising devices (e.g. two sheets of Polaroid) and place them in series in the path of an initially unpolarised beam of light. If they are perpendicular then we expect 'no' light to get through. However, if we put some test system between them which has the potential to rotate the polarisation of light, we will see this in the transmitted light. This kind of system is referred to as a polariser and analyzer, with the incident light falling on the polariser, passing through the test system, and then the analyzer. The intensity (proportional to field squared) goes as:

$$\propto 2A_1 = 2A_0 \cos 2\theta = I_0 \cos 2\theta$$

where, θ is the angle between the polarisers and I_o is the intensity of incident light (assuming no losses at all from the transmitted polarisation, which is never really the case).

Liquid Crystals

Liquid crystals are a particular group of materials which show a 'double' melting point: at the lower melting point the solid 'melts' but remains ordered in at least one dimension until some higher melting temperature is reached and the material becomes truly liquid. Between these two temperatures the material is called a liquid crystal. There are three types of liquid crystal: nematic, cholesteric and smectic, though only the first is widely used for its electrical properties. In the nematic class, the rod-shaped (or long ellipsoidal) molecules tend to line up parallel to each other locally — the extent of alignment may be enhanced by the application of relatively small electric fields. Lliquid crystals behave a bit like large permanent dipole fluid particles at low temperature (although the molecules are not permanently polarised). Nematic liquid crystals have the further property of aligning with respect to a true solid surface with which they are in contact with either perpendicular (homeotropic) or parallel (homogeneous) ordering.

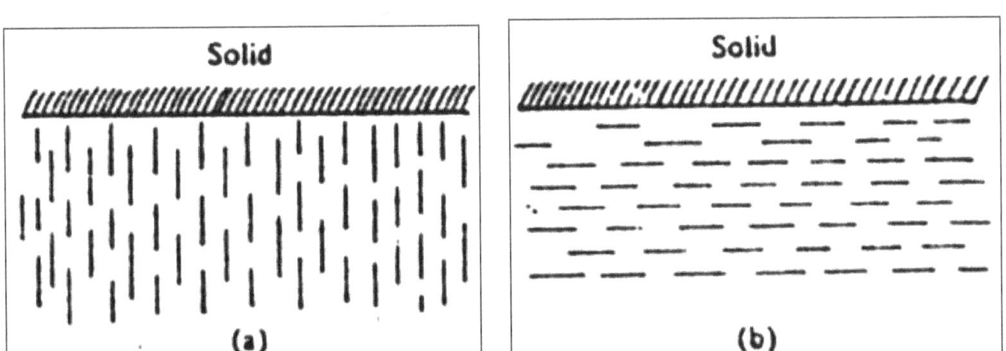

(a) Homeotropic and (b) homogeneous ordering of a nematic liquid crystal with respect to a solid surface.

If we take a thin cell (~10 μm) of homogeneously ordered nematic liquid crystal, as illustrated in figure then we can control the alignment through the application of an electric field E between the plates of the cell.

Although the required field 150 kV/m may seem high, this can be achieved with a relatively small voltage (1.5 V) across the 10 μm gap. The response time of the realignment is < 100 ms, which is adequate for most display purposes. This time dependence arises from Debye relaxation.

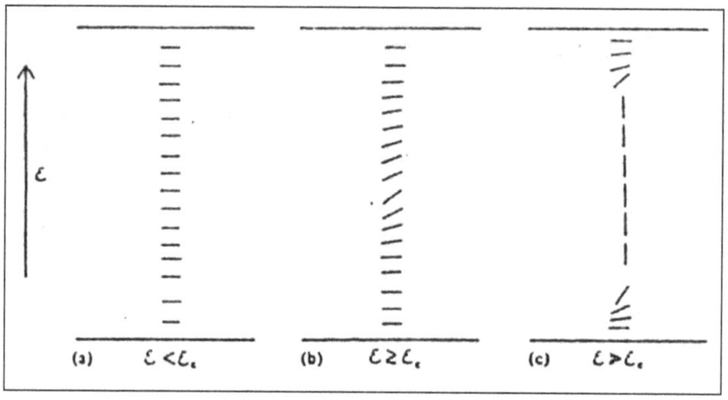

Realignment of a homogeneously ordered liquid crystal by an applied electric field, as might be used in liquid crystal displays.

Liquid Crystal Displays

Liquid crystal displays, or LCDs, have undergone a phenomenally rapid rate of development during the last 40 years. The first publications in this area appeared in 1968 where the authors were already hoping for applications in televisions, for example. However, early devices were unreliable and slow throughout the 1970s. In the 1980s they began their domination of the static/slow monochrome and colour display markets. In the late 1990s they became competitive for TV and computer displays. One principle drawback is the intensity — they are always present between the viewer and the light source and in reality always absorb some of the transmitted, even when 'on'. Furthermore some complain that the blacks are not competitive with other technologies, such as plasma TVs, because they are not perfectly absorbing when in the 'off' state. With the advent of LED backlighting in displays, it may be that the next 20 years sees a decline in the use of LCDs in favour of fast-switching sources (rather than fast-switching filters).

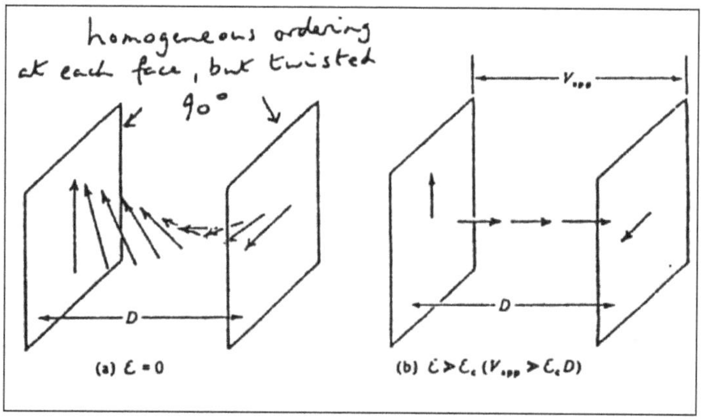

The principle of a twisted nematic display.

That said; let's look at basic principles of operation of the LCDs currently in use. A good example is the 90° twisted nematic display. The extra property used here is the ability of certain molecules to rotate the plane of polarisation of light, keeping the E-vector of the light aligned along the major axis of the molecule. There is only a very small change in angle from one molecule to the next. So, if the liquid crystal is held between two plates and then twisted by 90°, in the absence of any electric field, polarised light will follow the molecular axes and undergo a 90° rotation. If an electric field is now applied so as to alter the nematic alignment, the polarisation no longer rotates. Suitably arranged polariser/analyser combinations can thus distinguish between the two cases, and light can controllably be allowed to transmit through the device, or vice versa. A typical arrangement of the actual components is shown in figure, while other figure shows how this might be constructed in practice. The reflection mode is typically used in watches or pocket calculators, allowing them to work well around sources of bright ambient light (e.g. outside).

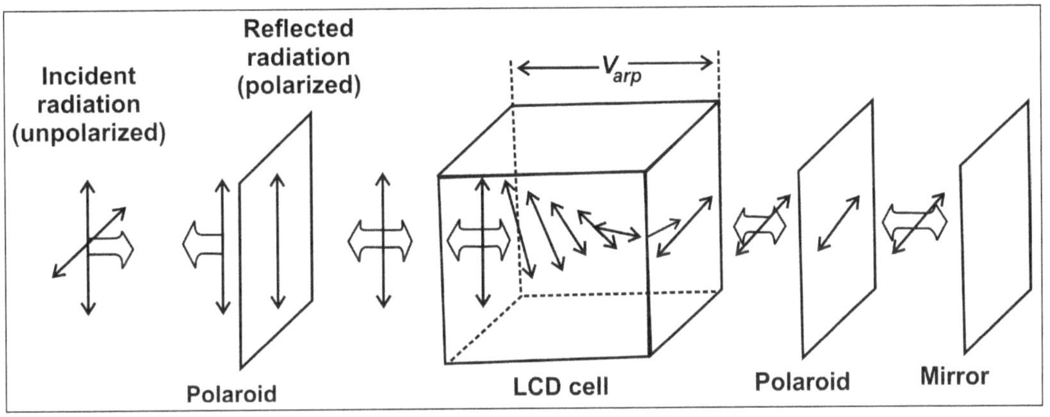

Light path in a reflection nematic display.

Magnetic Properties of Materials

The magnetic properties of a material are those which determine the ability of material to be suitable for a particular magnetic Application. Some of the typical magnetic properties of engineering materials are listed below:

- Permeability
- Retentivity or Magnetic Hysteresis
- Coercive force
- Reluctance

Permeability

It is the property of magnetic material which indicates that how easily the magnetic flux is build up in the material. Some time is also called as the magnetic susceptibility of material.

It is determined by the ratio of magnetic flux density to magnetizing force producing this magnetic flux density. It is denoted by μ.

Hence, $\mu = B/H$.

Where,

- B is the magnetic flux density in material in Wb/m^2.
- H is the magnetizing force of magnetic flux intensity in Wb/Henry-meter.

SI unit of magnetic permeability is Henry / meter.

Permeability of material is also defined as, $\mu = \mu_o \mu_r$

Where, μ_o is the permeability of air or vacuum, and $\mu_o = 4\pi \times 10^{-7}$ Henry/meter and μ_r is the relative permeability of material. $\mu_r = 1$ for air or vacuum.

A material selected for magnetic core in electrical machines should have high permeability, so that required magnetic flux can be produced in core by less ampere- turns.

Retentivity

When a magnetic material is placed in an external magnetic field, its grains get oriented in the direction of magnetic field. This results in magnetization of material in the direction of external magnetic field. Now, even after removal of external magnetic field, some magnetization exists, which is called residual magnetism. This property of material is called Magnetic retentively of material. A hysteresis loop or B-H cure of a typical magnetic material is shown in figure below.

Magnetization B_r in below hysteresis loop represents the residual magnetism of material.

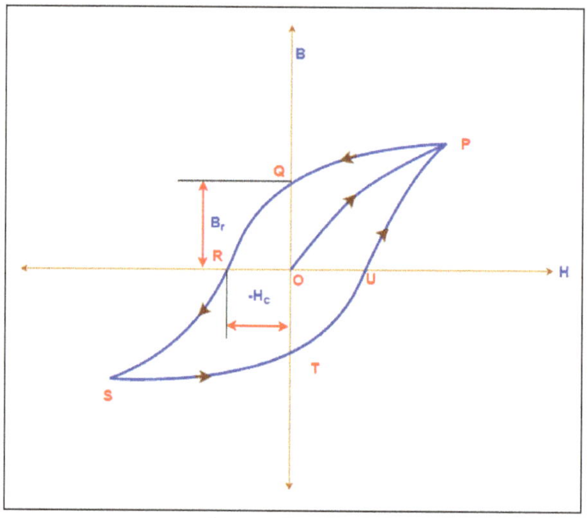

B-H Curve of a typical magnetic material.

Coercive Force

Due to retentivity of material, even after removal of external magnetic field some magnetization exists in material. This magnetism is called residual magnetism of material. To remove this residual magnetization, we have to apply some external magnetic field in opposite direction. This external magnetic motive force (ATs) required to overcome the residual magnetism is called “coercive force” of material. In above hysteresis loop, - H_c represents the coercive force.

The material having large value of residual magnetization and coercive force are called magnetically hard materials. The material having very low vale of residual magnetization and coercive force are called magnetically soft materials.

Reluctance

It is a property of magnetic material which resists to buildup of magnetic flux in material. It is denoted by R. Its unit is “Ampere-turns / Wb”.

Reluctance of magnetic material is given by:

$$\mathcal{R} = \frac{Ampure - Turns}{\phi}.$$

References

- Mechanical-properties: bu.edu.eg, Retrieved 21 March, 2019
- Chemical-properties-of-materials: electrical4u.com, Retrieved 25 May, 2019
- Electrical-properties-of-engineering-materials: electrical4u.com, Retrieved 15 July, 2019
- Thermal-properties-of-engineering-materials: engineeringenotes.com, Retrieved 5 April, 2019
- Producing-this-magnetic-flux-density, magnetic-properties-of-engineering-materials: electrical4u.com, Retrieved 29 June, 2019

4

Materials Bonding

Bonding in materials refers to the lasting attraction among atoms, molecules or ions which facilitates the creation of chemical compounds. They are broadly divided into primary bonds and secondary bonds. The topics elaborated in this chapter will help in gaining a better perspective about the different types of bonds such as ionic, covalent and hydrogen bonds.

Primary Bonds

Primary bonds involve sharing or donating electrons between atoms to form a more stable electron configuration. Place your mouse over the image below to see how electrons are shared between atoms.

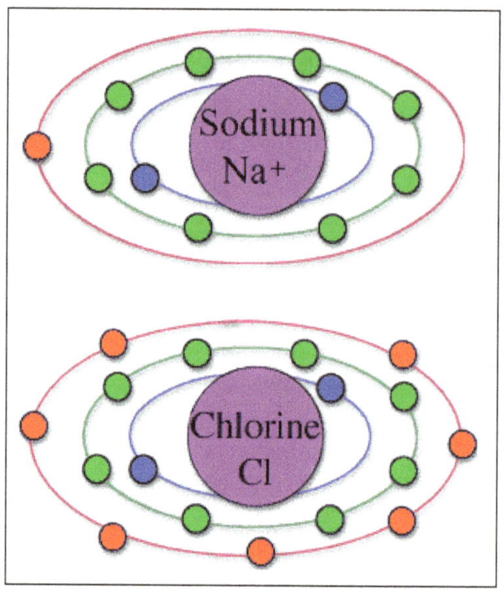

All elements except inert gases have an unfilled valence shell. For example, sodium has a nucleus containing 11 protons and orbiting shells containing 11 electrons. The outer shell has one valence electron.

Primary bonding occurs when electrons are lost or gained so that the outer shell is filled. For example, if a sodium atom loses its valence electron, it is left with a full outer shell of electrons and if a chlorine atom, which has only seven electrons in the outer shell, gains an electron, its outer shell is then full.

Ionic Bonding

An ionic bond can form between two elements when one has a small number of electrons in the valence shell (metal) and one has an almost full outer shell (non-metal). Atoms are more likely to accept electrons if they have an almost full outer shell (elements on the right of the periodic table).

H																	He
Li	Be											B	C	N	O	F	Ne
Na	Mg											Al	Si	P	S	Cl	Ar
K	Ca	Sc	Ti	V	Cr	Mn	Fe	Co	Ni	Cu	Zn	Ga	Ge	As	Se	Br	Kr
Rb	Sr	Y	Zr	Nb	Mo	Tc	Ru	Rh	Pd	Ag	Cd	In	Sn	Sb	Te	I	Xe
Cs	Ba	La*	Hf	Ta	W	Re	Os	Ir	Pt	Au	Hg	Tl	Pb	Bi	Po	At	Rn
Fr	Ra	Ac**															

RED = metals YELLOW = semi-conducting, or intermediate elements BLUE = nonmetals
* = Lanthanide or rare earth series ** = Actinide series

Sodium and chlorine form an ionic bond, with sodium giving up an electron from its valence shell and donating it to the chlorine atom to complete its valence shell.

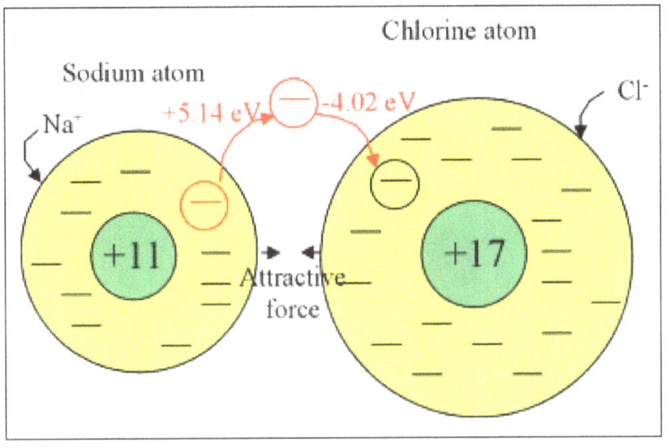

The non-metal atoms attract valence electrons from metal atoms to become negatively charged. Cooperatively, the metal atom becomes positively charged from losing one or more electrons.

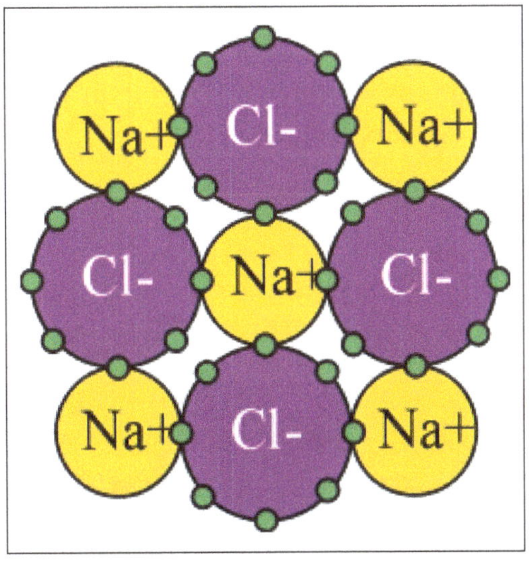

It is the attraction between two oppositely charged ions that forms the bond. This bond is non-directional, leading to freedom in the way they pack, but ions of opposite sign must surround each other to retain the attraction between ions.

Ionic bonds are strong and stiff. As a result they generally give a material with:

- High strength
- High elastic modulus
- High melting point
- Poor electrical conductivity

Some examples of ionic bonding are:

- Magnesia (MgO)
- Alumina (Al_2O_3)

Covalent Bonding

In covalent bonding, a stable electron configuration is created by sharing of electrons between neighbouring atoms. Two atoms that are covalently bonded will share at least one electron from each atom.

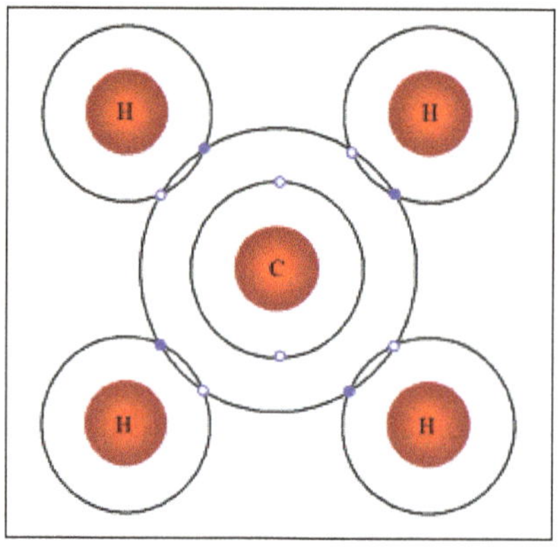

Covalent bonding in methane.

The electrons are shared between atoms within in a 3D structure where the bonds are highly directional. This directionality dictates the atomic packing.

In diamond, the carbon atoms build up into a 3D array with the bonds pointing towards the corners of a tetrahedron.

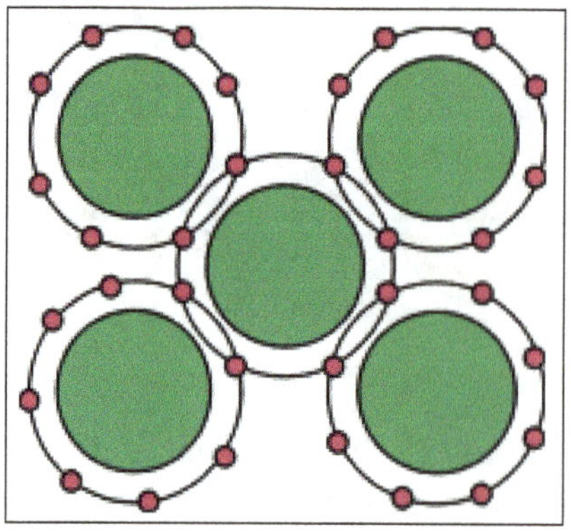

Diamond atoms sharing electrons.

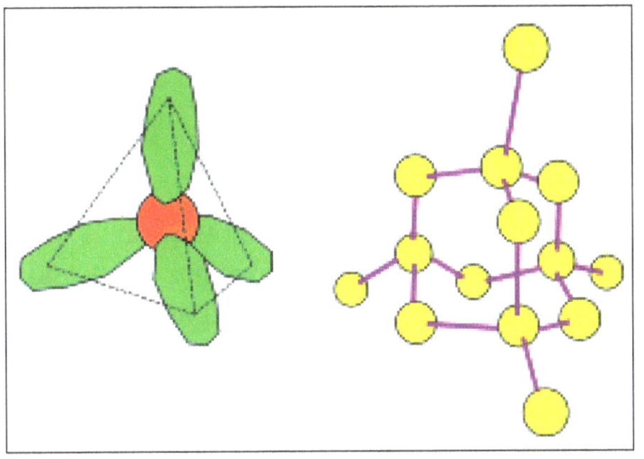

Covalent bonds can be very stiff and generally give a material with a:

- Very high elastic modulus
- High (inherent) strength
- High melting point
- Low electrical conductivity

Covalent bonding is the dominant bonding found in silicate ceramics and glasses. It also occurs in the backbone of polymer chains and in the cross-links in thermosetting polymers.

Types of Covalent Bonds

The bonds formed by sharing of an electron pair between two atoms can typically be single, double or triple; that is, in covalent bonds, atoms may be linked together by single, double or triple bonds. However there is also quadruple covalent bond. In simplest terms, "single", "double", and "triple" refer to the number of shared electron pairs in the bond. For example, in the dioxygen (O_2) molecule, each oxygen must share two of its electrons in order to obtain the noble gas configuration of 8 valence electrons, resulting in a double bond of two shared electron pairs. The *valency* of an atom is the maximum number of bonds it can form, usually the number of electrons required to reach a stable noble gas configuration (there are exceptions to this rule, called the *octet rule* as all noble gases except helium have 8 valence electrons. Each double bond counts as 2 bonds for this purpose, and each triple bond counts as 3.

Whether a bond is single, double, or triple is its *bond order*.

Single Covalent Bond

In some molecules, shared pair of electrons is contributed between the atoms, thus creating a "single bond". For example, in a hydrogen molecule, the two atoms of hydrogen are bonded together by a single bond.

How? The *valency* of hydrogen is 1, that is, the number of electrons present in the outermost shell is 1 and it needs 1 more electron to attain a noble gas configuration, that is a completely filled outer shell.

H has a valency of 1 and it needs 1 more electron to become stable. So, it forms a bond with another hydrogen atom and shares its electron with it and the second hydrogen atom shares its 1 electron with the first one,

(Hx) (xH)

(Here, x shows the number of electrons present in the outer shell of a hydrogen atom.) Now, the first atom shares its one electron with the second hydrogen atom and the second atom shares its one electron with the first one and thus we get one hydrogen atom,

(H(xx)H)

Thus, 2 atoms of hydrogen share their electron to form a molecule of hydrogen, H_2. This allows each hydrogen atom to attain the electronic configuration of the noble gas helium. The shared pair of electrons is said to constitute a single bond within the two hydrogen atoms. It is a single bond because only one pair of electrons is shared within the hydrogen atoms. A single bond is denoted by a single line between the atoms. For example, the covalent bond in a hydrogen molecule is represented by:

H—H

Such single covalent bond is also formed in chlorine molecule, Cl_2.

Double Covalent Bonds

Sometimes, two atoms share two, not just one, electron pairs to attain a noble gas configuration, with each atom contributing two electrons for a total of four bonding electrons. The minimum valency of an atom that can participate in this kind of covalent bonding must have a minimum valence of 2. O has a valency of 2, with 6 electrons, needing 2 more to become stable (although oxygen has 6 electrons, it has a valency of 2 because it only needs 2 more for stability). It therefore can form a *double bond* with another oxygen atom, with each atom keeping 4 of its electrons uninvolved in bonding and sharing 2 with the other atom.

(xxxxOxx) (xxOxxxx)

thus becomes:

(xxxxO(xxxx)Oxxxx)

4 kept electrons plus 4 shared electrons then equal a stable 8-electron configuration. A double bond is denoted by a doubled line between the bonding atoms:

O=O

Such a double bond is also found in the carbon dioxide molecule CO_2:

(xxxxO(xxxx)C(xxxx)Oxxxx)

Here, each oxygen atom double-bonds to the carbon atom. The carbon atom has 4 valence electrons, and since it needs 4 more, it has a valency of 4, allowing it to form 2 double bonds (or 4 total bonds):

O=C=O

Triple Covalent Bonds

Likewise, when two atoms share three electron pairs, the bonding interaction is a *triple bond*. The minimum valency of an atom that can participate in this kind of bonding must be a minimum of 3.

N has a valency of 3, with 5 electrons, needing 3 more to become stable. It can form a triple bond with another nitrogen atom, with each atom sharing a total of 6 electrons with 2 unshared, making a full octet,

(xxNxxx) (xxxNxx)

bonds to form:

(xxN(xxxxxx)Nxx)

A triple bond is denoted by three lines connecting the bonding atoms:

$$\underset{109.76\text{pm}}{\underleftrightarrow{N \equiv N}}$$

Metallic Bonding

In metallic bonding, the electrons are "surrendered" to a common pool and become shared by all the atoms in the solid metal.

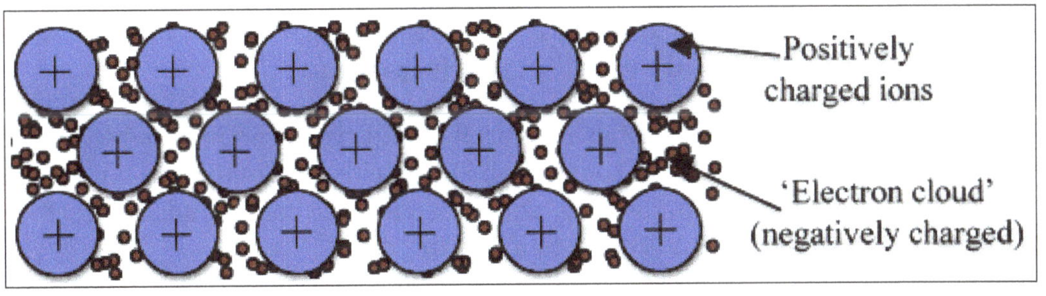

Metallic bonding is found in metals and their alloys. When the atoms give up their valence electrons, they form ions. These ions are held together by the electron cloud surrounding them.

Metallic bonding generally results in a material being strong and stiff and gives:

- High elastic modulus;
- High strength;
- Good electrical conductivity (because the electrons can move easily);
- Metallic lustre;
- High ductility.

The outer energy levels of metal atoms (the *s* and *p* orbitals) overlap. At least one of the valence electrons participating in a metallic bond is not shared with a neighbor atom, nor is it lost to form an ion. Instead, the electrons form what may be termed an "electron sea" in which valence electrons are free to move from one atom to another.

The electron sea model is an oversimplification of metallic bonding. Calculations based on electronic band structure or density functions are more accurate. Metallic bonding may be seen as a consequence of a material having many more delocalized energy states than it has delocalized electrons (electron deficiency), so localized unpaired electrons may become delocalized and mobile. The electrons can change energy states and move throughout a lattice in any direction.

Bonding can also take the form of metallic cluster formation, in which delocalized electrons flow around localized cores. Bond formation depends heavily on conditions. For example, hydrogen is a metal under high pressure. As pressure is reduced, bonding changes from metallic to nonpolar covalent.

Relating Metallic Bonds to Metallic Properties

Because electrons are delocalized around positively charged nuclei, metallic bonding explains many properties of metals.

- Electrical conductivity: Most metals are excellent electrical conductors because the electrons in the electron sea are free to move and carry charge. Conductive non-metals (such as graphite), molten ionic compounds, and aqueous ionic compounds conduct electricity for the same reason-electrons are free to move around.
- Thermal conductivity: Metals conduct heat because the free electrons are able to transfer energy away from the heat source and also because vibrations of atoms (phonons) move through a solid metal as a wave.
- Ductility: Metals tend to be ductile or able to be drawn into thin wires because local bonds between atoms can be easily broken and also reformed. Single atoms or entire sheets of them can slide past each other and reform bonds.
- Malleability: Metals are often malleable or capable of being molded or pounded into a shape, again because bonds between atoms readily break and reform. The binding force between metals is non-directional, so drawing or shaping a metal is less likely to fracture it. Electrons in a crystal may be replaced by others. Further, because the electrons are free to move away from each other, working a metal doesn't force together like-charged ions, which could fracture a crystal through the strong repulsion.
- Metallic luster: Metals tend to be shiny or display metallic luster. They are opaque once a certain minimum thickness is achieved. The electron sea reflects photons off the smooth surface. There is an upper-frequency limit to the light that can be reflected.

The strong attraction between atoms in metallic bonds makes metals strong and gives them high density, high melting point, high boiling point, and low volatility. There are exceptions. For example, mercury is a liquid under ordinary conditions and has a high vapor pressure. In fact, all of the metals in the zinc group (Zn, Cd, and Hg) are relatively volatile.

Secondary Bonds

Secondary bonds are weak in comparison to primary bonds. They are found in most materials, but their effects are often overshadowed by the strength of the primary bonding.

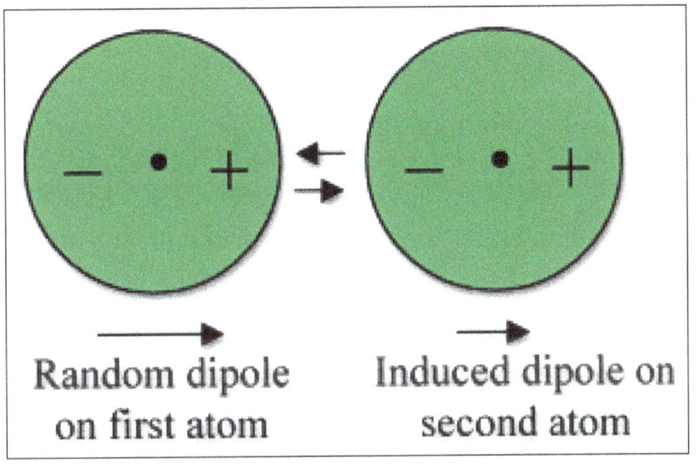

Secondary bonding occurs due to forces produced by atomic or molecular dipoles. The dipoles are produced by random fluctuation of the electrons around electrically symmetric atoms. Once a random dipole is formed in one atom, a dipole is induced in the adjacent atom. These dipoles cause a weak attraction to form between the oppositely charged ends of the dipoles. This is the type of bonding present in N_2 molecules, and is known as van der Waals Bonding.

Another case of secondary bonding is caused when there exists a permanent dipole in a molecule due to an asymmetrical arrangement of positive and negative regions. Molecules with a permanent dipole can both induce a dipole in adjacent electrically symmetrical molecules (and thus form a weak bond), or can form bonds with other permanent dipole molecules.

Van Der Waals Bonding

Van der Waals Forces are the weak forces which contribute to intermolecular bonding between molecules. Molecules inherently possess energy and their electrons are always in motion, so transient concentrations of electrons in one region or other lead electrically positive regions of a molecule to be attracted to the electrons of another molecule. Similarly, negatively-charged regions of one molecule are repulsed by negatively-charged regions of another molecule.

Van der Waals forces are the sum of the attractive and repulsive electrical forces between atoms and molecules. These forces differ from covalent and ionic chemical bonding because they result from fluctuations in charge density of particles. Examples of van der Waals forces include hydrogen bonding, dispersion forces, and dipole-dipole interactions.

Van der Waals Forces

- Van der Waals forces are distance-dependent forces between atoms and molecule not associated with covalent or ionic chemical bonds.
- Sometimes the term is used to encompass all intermolecular forces, although some scientists only include London dispersion force, Debye force, and Keeson force.
- Van der Waals forces are the weakest of the chemical forces, but they still play an important role in the properties of molecules and in surface science.

Properties of Van der Waals Forces

Certain characteristics are displayed by van der Waals forces:

- The forces are additive.
- The forces are weaker than either ionic or covalent chemical bonds.
- They are not directional.
- The forces act only over a very short range. The interaction is greater when molecules draw closer.
- Van der Waals forces are independent of temperature, with the exception of dipole-dipole interactions.

Components of Van der Waals Forces

Van der Waals forces are the weakest intermolecular forces. Their strength typically ranges from 0.4 to 4 kJ/mol and act over distances less than 0.6 nm. When the distance is less than 0.4 nm, the net effect of the forces is repulsive as electron clouds repel each other.

There are four major contributions to van der Waals forces:

- There is a negative component that prevents molecules from collapsing. This is due to the Pauli Exclusion Principle.
- Either an attractive or repulsive electrostatic interaction occurs between permanent charges, dipoles, quadrupoles, and multipoles. This interaction is called Keesom interaction or Keesom force, named for Willem Hendrik Keesom.
- Induction or polarization occurs. This is an attractive force between a permanent polarity on one molecule and an induced polarity on another. This interaction is called the Debye force for Peter J.W. Debye.
- London dispersion force is the attraction between any pair of molecules due to instantaneous polarization. The force is named after Fritz London. Note even nonpolar molecules experience London dispersion.

Causes of Van der Waals Forces

Quantum Mechanics strongly emphasizes the constant movement of electrons in an atom through the Schrödinger Equation and the Heisenberg's Uncertainty Principle. The Heisenberg's Uncertainty Principle proposes that the energy of the electron is never zero; therefore, it is constantly moving around its orbital. The square of the Schrödinger Equation for a particle in a box suggests that it is probable of finding the electron (particle) anywhere in the orbital of the atom.

These two important aspects of Quantum Mechanics strongly suggest that the electrons are constantly are moving in an atom, so dipoles are probable of occurring. A dipole is defined as molecules or atoms with equal and opposite electrical charges separated by a small distance.

It is probable to find the electrons in this state:

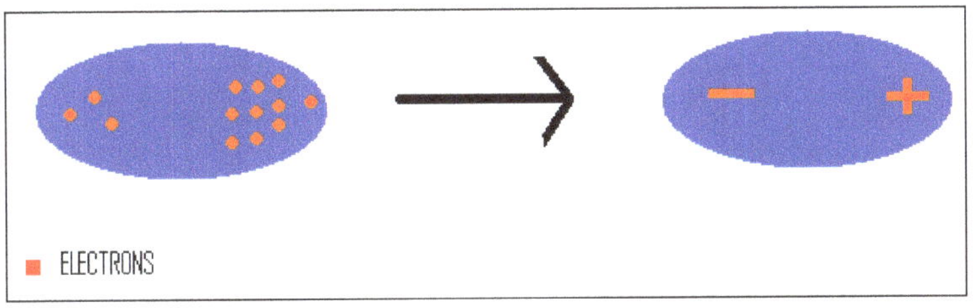

This is how spontaneous (or instantaneous) dipoles occur. When groups of electrons move to one end of the atom, it creates a dipole. These groups of electrons are constantly moving so they move from one end of the atom to the other and back again continuously. Therefore, the opposite state is as probable of occurring.

Opposite state due to fluctuation of dipoles:

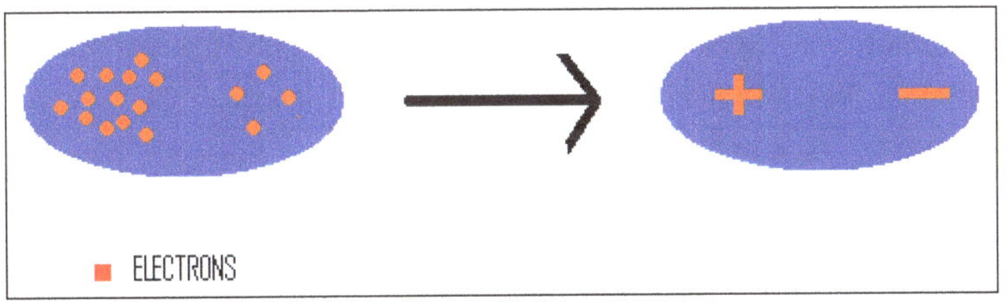

Dipole-Dipole Interaction

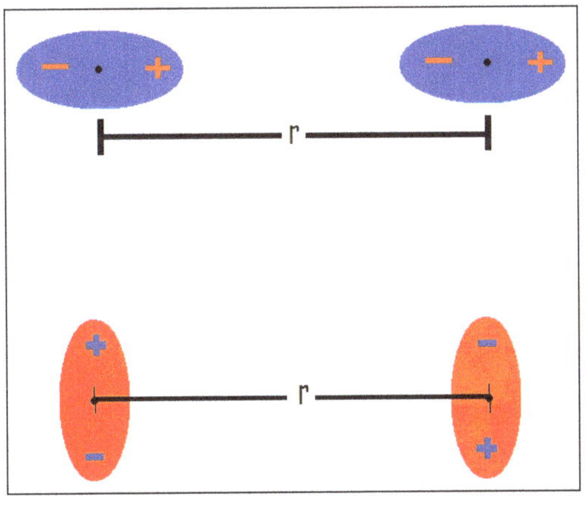

Dipole-Dipole interactions occur between molecules that have permanent dipoles; these molecules are also referred to as polar molecules. The figure shows the electrostatic interaction between two dipoles.

The potential energy of the interaction for the top pair of the image above is represented by the equation:

$$V = -\frac{2\mu_A \mu_B}{4\pi\epsilon_o r^3}$$

The potential energy of the interaction for the bottom pair is represented by the equation:

$$V = -\frac{\mu_A \mu_B}{4\pi\epsilon_o r^3}$$

with,

- V is the potential energy.
- μ is the dipole moment.
- ϵ_0 is the vacuum permittivity.
- r is the length between the two nuclei.

The negative sign indicates that energy is released out of the system, because energy is released when bonds are formed, even weak bonds. The negative sign also suggests that the interaction is driven by an attractive force (a positive sign would indicate a repulsion force between the two molecules). If the conditions of these two samples are the same except for their orientation, the second pair of the electron will always have a larger potential energy, because both the negative and positive ends are involve in the interactions.

Induced Dipoles

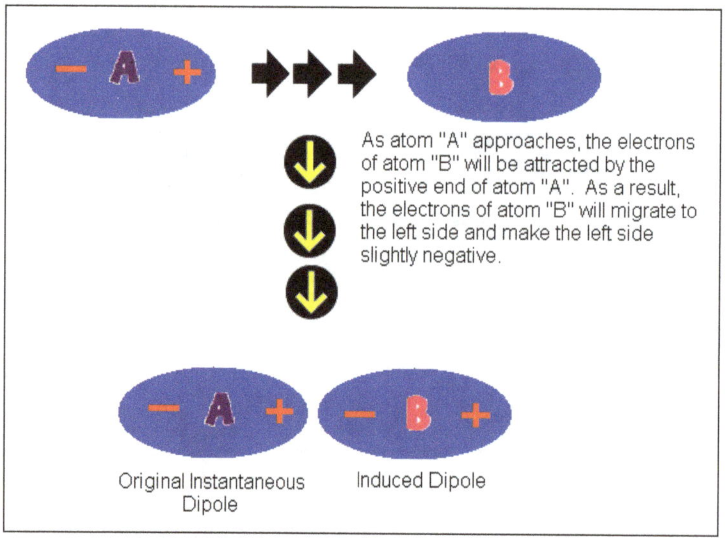

An induced dipole moment is a temporary condition during which a neutral nonpolar atom (i.e. Helium) undergoes a separation of charges due to the environment. When an instantaneous dipole atom approaches a neighboring atom, it can cause that atom to also produce dipoles. The neighboring atom is then considered to have an induced dipole moment.

Even though these two atoms are interacting with each other, their dipoles may still fluctuate. However, they must fluctuate in synchrony in order to maintain their dipoles and stay interacted with each other. Result of synchronizing fluctuation of dipoles:

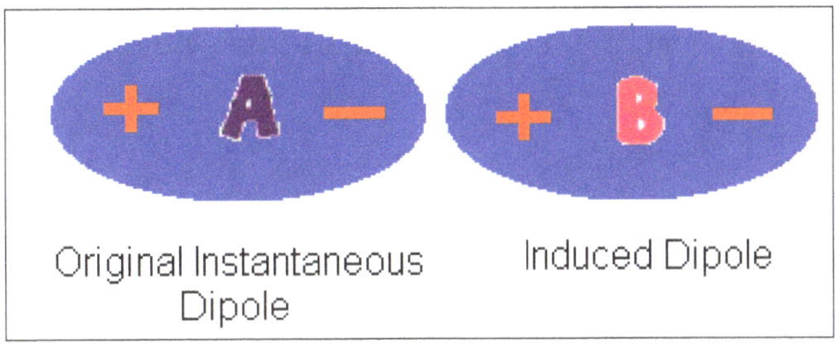

The potential energy representing the dipole-induced dipole interaction is:

$$V = -\frac{\alpha\mu^2}{4\pi\epsilon_o r^6}$$

where, α = polarizability of the nonpolar molecule.

Polarizability defines how easy the electron density of an atom or a molecule can be distorted by an external electric field.

Spontaneous Dipole-Induced Dipole Interaction

Spontaneous dipole-induced dipole interactions are also known as dispersion or London forces (name after the German physicist Fritz London). They are large networks of intermolecular forces between nonpolar and non-charged molecules and atoms (i.e. alkanes, noble gases, and halogens). Molecules that have induced dipoles may also induce neighboring molecules to have dipole moments, so a large network of induced dipole-induced dipole interactions may exist. The image below illustrates a network of induced dipole-induced dipole interactions.

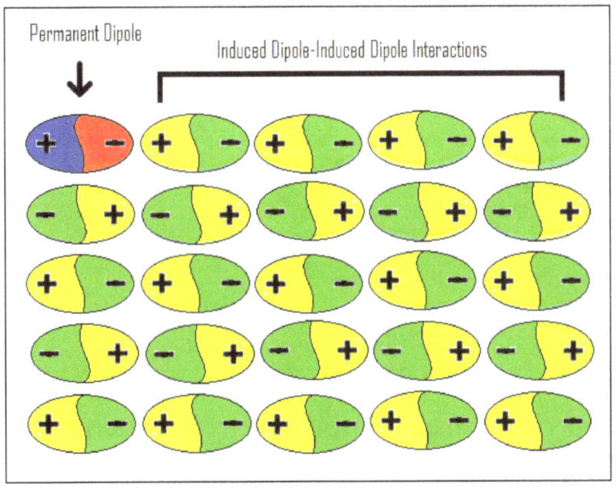

The potential energy of an induced dipole-induced dipole interaction is represented by this equation:

$$V = -\frac{3}{2}\frac{I_a I_b}{I_a + I_b}\frac{\alpha_a \alpha_b}{r^6}$$

where, I = The first ionization energy of the molecule

The radius is a huge determinant of how large the potential energy is since the potential energy is inversely proportional to r^6. A small increase in the radius would greatly decrease the potential energy of the interaction.

Hydrogen Bonding

Hydrogen bonding is the strongest form of secondary bonding and is formed from the polar nature of molecules containing hydrogen.

The hydrogen side of the molecule is more positive than the atom it is bonded to, allowing an attraction to form with the negative end of another molecule.

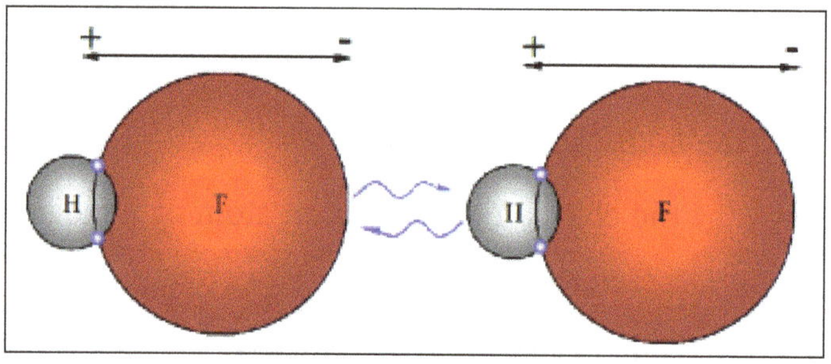

Hydrogen bond in HF.

In hydrogen bonding, the hydrogen is attached directly to an element which is electronegative. This causes the hydrogen to acquire a significant positive charge.

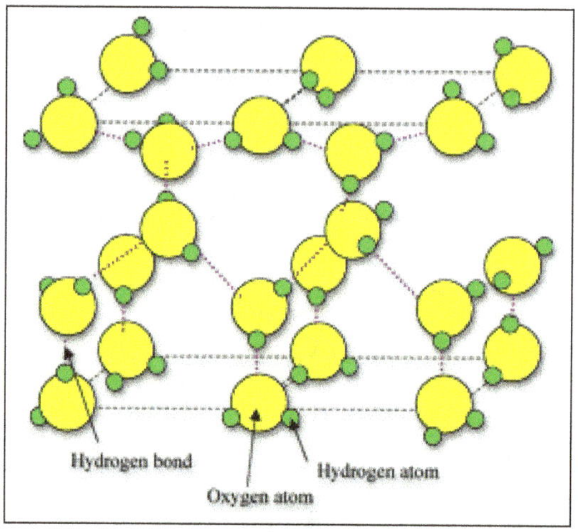

The melting and boiling points of materials containing hydrogen bonding, are abnormally high, for their atomic weights, due to hydrogen bonding.

Types of Hydrogen Bonds

Hydrogen bonds can occur within one single molecule, between two like molecules, or between two unlike molecules

- Intramolecular hydrogen bonds: Intramolecular hydrogen bonds are those which occur within one single molecule. This occurs when two functional groups of a molecule can form hydrogen bonds with each other. In order for this to happen, both a hydrogen donor an acceptor must be present within one molecule, and they must be within close proximity of each other in the molecule. For example, intramolecular hydrogen bonding occurs in ethylene glycol ($C_2H_4(OH)_2$) between its two hydroxyl groups due to the molecular geometry.

- Intermolecular hydrogen bonds: Intermolecular hydrogen bonds occur between separate molecules in a substance. They can occur between any numbers of like or unlike molecules as long as hydrogen donors and acceptors are present in positions in which they can interact. For example, intermolecular hydrogen bonds can occur between NH_3 molecules alone, between H_2O molecules alone, or between NH_3 and H_2O molecules.

Properties and Effects of Hydrogen Bonds

When we consider the boiling points of molecules, we usually expect molecules with larger molar masses to have higher normal boiling points than molecules with smaller molar masses. This, without taking hydrogen bonds into account, is due to greater dispersion forces. Larger molecules have more space for electron distribution and thus more possibilities for an instantaneous dipole moment. However, when we consider the table below, we see that this is not always the case.

We see that H_2O, HF, and NH_3 each have higher boiling points than the same compound formed between hydrogen and the next element moving down its respective group, indicating that the former have greater intermolecular forces. This is because H_2O, HF, and NH_3 all exhibit hydrogen bonding, whereas the others do not. Furthermore, H_2O has a smaller molar mass than HFHF, but partakes in more hydrogen bonds per molecule, so its boiling point is consequently higher.

Viscosity

The same effect that is seen on boiling point as a result of hydrogen bonding can also be observed in the viscosity of certain substances. Those substances which are capable of forming hydrogen bonds tend to have a higher viscosity than those that do not. Substances which have the possibility for multiple hydrogen bonds exhibit even higher viscosities.

Factors Preventing Hydrogen Bonding

- Electronegativity: Hydrogen bonding cannot occur without significant electronegativity differences between hydrogen and the atom it is bonded to. Thus, we see molecules such as PH_3, which no not partake in hydrogen bonding. PH_3 exhibits a trigonal pyramidal molecular geometry like that of ammmonia, but unlike NH_3 it cannot hydrogen bond. This is due to the similarity in the electro-negativities of phosphorous and hydrogen. Both atoms have an electronegativity of 2.1, and thus, no dipole moment occurs. This prevents the hydrogen bonding from acquiring the partial positive charge needed to hydrogen bond with the lone electron pair in another molecule.

- Atom Size: The size of donors and acceptors can also affect the ability to hydrogen bond. This can account for the relatively low ability of Cl to form hydrogen bonds. When the radii of two atoms differ greatly or are large, their nuclei cannot achieve close proximity when they interact, resulting in a weak interaction.

Hydrogen Bonding in Nature

Hydrogen bonding plays a crucial role in many biological processes and can account for many natural phenomena such as the unusual properties of water. In addition to being present in water, hydrogen bonding is also important in the water transport system of plants, secondary and tertiary protein structure, and DNA base pairing.

Plants

The cohesion-adhesion theory of transport in vascular plants uses hydrogen bonding to explain many key components of water movement through the plant's xylem and other vessels. Within a vessel, water molecules hydrogen bond not only to each other, but also to the cellulose chain which comprises the wall of plant cells. This creates a sort of capillary tube which allows for capillary action to occur since the vessel is relatively small. This mechanism allows plants to pull water up into their roots. Furthermore, hydrogen bonding can create a long chain of water molecules which can overcome the force of gravity and travel up to the high altitudes of leaves.

Proteins

Hydrogen bonding is present abundantly in the secondary structure of proteins, and also sparingly in tertiary conformation. The secondary structure of a protein involves interactions (mainly hydrogen bonds) between neighboring polypeptide backbones which contain Nitrogen-Hydrogen bonded pairs and oxygen atoms. Since both N and O are strongly electronegative, the hydrogen atoms bonded to nitrogen in one polypeptide backbone can hydrogen bond to the oxygen atoms in another chain and visa-versa. Though they are relatively weak, these bonds offer great stability to secondary protein structure because they repeat a great number of times.

In tertiary protein structure, interactions are primarily between functional R groups of a polypeptide chain; one such interaction is called a hydrophobic interaction. These interactions occur because of hydrogen bonding between water molecules around the hydrophobe and further reinforce conformation.

References

- Primary-bonds: materials.unsw.edu.au, Retrieved 2 August, 2019
- Ionic-bonding: materials.unsw.edu.au, Retrieved 22 April, 2019
- Covalent-bonding: materials.unsw.edu.au, Retrieved 27 June, 2019
- Metallic-bonding: unsw.edu.au, Retrieved 26 January, 2019

- Metallic-bond-definition-properties-and-examples: thoughtco.com, Retrieved 23 August, 2019
- Definition-of-van-der-waals-forces: thoughtco.com, Retrieved 10 May, 2019
- Van-Der-Waals-Interactions, Specific-Interactions, Atomic-and-Molecular-Properties, Physical-and-Theoretical-Chemistry: libretexts.org, Retrieved 19 March, 2019
- Hydrogen-bonding: materials.unsw.edu.au, Retrieved 11 July, 2019
- Hydrogen-Bonding, Intermolecular-Forces, Atomic-and-Molecular-Properties, Physical-and-Theoretical-Chemistry: libretexts.org, Retrieved 9 February, 2019

5

Sub-fields of Materials Science

Material science is an interdisciplinary field which has numerous sub-fields. A few of them are ceramic engineering, metallurgy, polymer science and engineering, and nanotechnology. This chapter has been carefully written to provide an easy understanding of these sub-fields of materials science.

Ceramic Engineering

Ceramic engineering is the technology that involves the design and manufacture of ceramic products. Ceramics are inorganic, non-metallic materials that have been hardened by baking at high temperatures. Highly regarded for being resistant to heat, these materials can be used for many demanding tasks that other materials, such as metals and polymers, cannot.

Until about the mid-twentieth century, the most important ceramics were the traditional clays, which were made into pottery, dinnerware, bricks, tiles, and decorative items. Since then, new materials called advanced ceramics have been prepared and are being used for a wide range of applications, including components used by the aerospace, automotive, defense, environmental, fiber-optic, and medical technologies. Ceramic parts are also used in cellular phones and personal computers. In addition, ceramic engineering is useful in such areas as petroleum refining, food processing, packaging, and mining.

The space shuttle has a coat of ceramic tiles that protect it from the searing heat produced during reentry into the atmosphere.

Traditional ceramic raw materials include clay minerals such as kaolinite. Modern ceramic materials include silicon carbide and tungsten carbide, both of which are highly resistant to abrasion and are used in applications such as the wear plates of crushing equipment in mining operations. Each of NASA's Space Shuttles has a coating of ceramic tiles that protect it from the searing heat (up to 2,300 °F) produced during re-entry into Earth's atmosphere. Thus, ceramic engineering is an important contributor to the modern technological revolution.

Ceramic Process

The ceramic process generally follows this flow.

Milling → Batching → Mixing → Forming → Drying → Firing → Assembly

Milling is the process by which materials are reduced from a larger size to a smaller size. Milling may involve breaking up cemented material, thus the individual particle retain their shape or pulverization which involves grinding the particles themselves to a smaller size. Pulverization is actually fracturing the grains and breaking them down.

Generally, milling is done through mechanical means. The means include attrition which is particle to particle collision that results in agglomerate break up or particle shearing. Compression which is applying compressive forces that result in break up or fracturing. Another means is impact which involves a milling media—or the particles themselves—that cause break up or fracturing.

Examples of equipment that achieve attrition milling is a planetary mill or an wet attrition mill, also called wet scrubber. A wet scrubber is a machine that has paddles in water turning in opposite direction causing two vortexes turning into each other. The material in the vortex collides and breaks up.

Equipment that achieves compression milling includes a jaw crusher, roller crusher, and cone crushers. Finally, impact mills may include a ball mill with media that tumble and fracture material. Shaft impactors cause particle to particle attrition and compression which achieve size reduction.

- Batching is the process of weighing the oxides according to recipes, and preparing them for mixing and drying.
- Mixing occurs after batching and involves a variety of equipment such as dry mixing ribbon mixers (a type of cement mixer), Mueller mixers, and pug mills. Wet mixing generally involves the same equipment.
- Forming is making the mixed material into shapes, ranging from toilet bowls to spark plug insulators. Forming can involve: Extrusion, such as extruding "slugs" to make bricks, pressing to make shaped parts, or slip casting, as in making toilet bowls, wash basins and ornamentals like ceramic statues.
- Forming produces a "green" part, ready for drying. Green parts are soft, pliable, and over time will lose shape. Handling the green product will change its shape. For example, a green brick can be "squeezed," and after squeezing it will stay that way.
- Drying is removing the water or binder from the formed material. Spray drying is widely used to prepare powder for pressing operations. Other dryers are tunnel dryers and

periodic dryers. Controlled heat is applied in this two-stage process. First, heat removes water. This step needs careful control, as rapid heating cause's cracks and surface defects. The dried part is smaller than the green part, and is brittle, necessitating careful handling, since a small impact will cause crumbling and breaking.

- Firing is where the dried parts pass through a controlled heating process, and the oxides are chemically changed to cause sintering and bonding. The fired part will be smaller than the dried part.
- Assembly: This process is for parts that require additional subassembly parts. In the case of a spark plug, the electrode is put into the insulator. This step does not apply to all ceramic products.

Applications

Ceramics can be used for many technological applications. One example is NASA's Space Shuttle, which uses ceramic tiles to protect it from the searing heat of re-entry into Earth's atmosphere. Future supersonic space planes may likewise be fitted with such tiles. Ceramics are also used widely in electronics and optics. In addition to the applications listed here, ceramics are used as a coating in various engineering cases. For example, a ceramic bearing coating may be used over a titanium frame for an airplane. Recently, the field has come to include the studies of single crystals or glass fibers, in addition to traditional polycrystalline materials, and the applications of these have been overlapping and changing rapidly.

Aerospace

- Engines; shielding a hot running airplane engine from damaging other components.
- Airframes; Used as a high-stress, high-temp and lightweight bearing and structural component.
- Missile nose-cones; shielding the missile internals from heat.
- Space Shuttle tiles.
- Rocket Nozzles; Withstands and focuses the exhaust of the rocket booster.

Biomedical

- Artificial bone; Dentistry applications, teeth.
- Biodegradable splints; Reinforcing bones recovering from osteoporosis.
- Implant material.

Electronics and Electrical Industry

- Capacitors
- Integrated Circuit packages

- Transducers
- Insulators

Optical/Photonic

- Optical fibers; Glass fibers for super-fast data transmission
- Switches
- Laser amplifiers
- Lenses

Modern-day Ceramic Engineering

Modern-day ceramic engineers may find themselves in a variety of industries, including mining and mineral processing, pharmaceuticals, foods, and chemical operations.

Now a multi-billion dollar a year industry, ceramics engineering and research has established itself as an important field of science. Applications continue to expand as researchers develop new kinds of ceramics to serve different purposes. An incredible number of ceramics engineering products have made their way into modern life.

Metallurgy

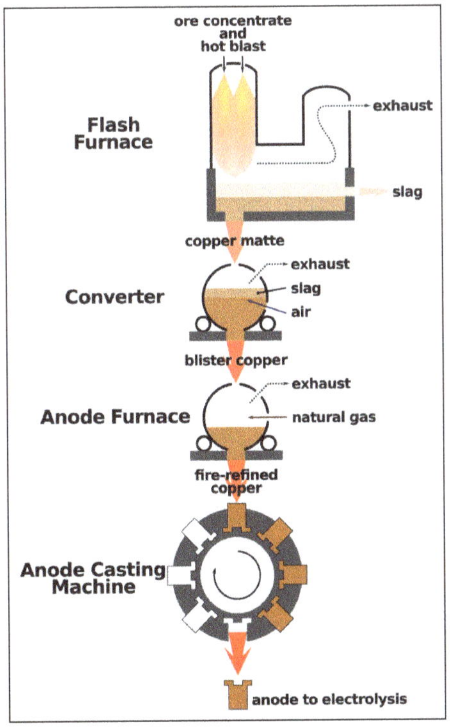

Copper Flash Smelting Process.

Metallurgy is defined as a process that is used for the extraction of metals in their pure form. The compounds of metals mixed with soil, limestone, sand, and rocks are known as minerals. Metals are commercially extracted from minerals at low cost and minimum effort. These minerals are known as ores. A substance which is added to the charge in the furnace to remove the gangue (impurities) is known as flux. Metallurgy deals with the process of purification of metals and the formation of alloys.

Principles of Metallurgy

The metallurgical process can be classified as the following:

1. Crushing and grinding: The first process in metallurgy is crushing of ores into a fine powder in a crusher or ball mill. This process is known as pulverization.
2. The concentration of ores: The process of removing impurities from ore is known as a concentration of minerals or ore dressing. In metallurgy, we concentrate the ores mainly by the following methods.
3. Hydrolytic method: In this method, we pour the ore over a sloping, vibrating corrugated table with grooves. A jet of water is allowed to flow over the surface. The denser ore particles settle in the grooves, and the impurities are washed away by water.
4. Magnetic separation: In this case, the crushed ore is placed on a conveyor belt. This belt rotates around two wheels in which one of the wheels is magnetic, and therefore the magnetic particles get attracted to the magnetic wheel and fall apart from the non-magnetic particles.
5. Froth floatation: In this process, we take the crushed ore in a large tank which contains oil and water. A current of compressed air is passed through it. The ore gets wet by oil and is separated from the impurities in the form of froth. Ore is lighter, and so it comes on the surface and impurities are left behind.
6. Roasting and calcination: In metallurgy, the process of heating a concentrated ore in the presence of oxygen is known as roasting. This process is applied in a case of sulfide ores. For ores containing carbonate or hydrated oxides, heating is done in the absence of air to melt the ores, and this process is known as calcination.

Extractive Metallurgy

Following separation and concentration by mineral processing, metallic minerals are subjected to extractive metallurgy, in which their metallic elements are extracted from chemical compound form and refined of impurities.

Metallic compounds are frequently rather complex mixtures (those treated commercially are for the most part sulfides, oxides, carbonates, arsenides, or silicates), and they are not often types that permit extraction of the metal by simple, economical processes. Consequently, before extractive metallurgy can affect the separation of metallic elements from the other constituents of a compound, it must often convert the compound into a type that can be more readily treated. Common practice is to convert metallic sulfides to oxides, sulfates, or chlorides; oxides to sulfates or chlorides; and carbonates to oxides. The processes that accomplish all this can be categorized as either pyrometallurgy or hydrometallurgy. Pyrometallurgy involves heating operations such as roasting, in which compounds are converted at temperatures just below their melting points, and smelting, in which all the constituents of an ore or concentrate are completely melted and separated into two liquid layers, one containing the valuable metals and the other the waste rock. Hydrometallurgy consists of such operations as leaching, in which metallic compounds are selectively dissolved from an ore by an aqueous solvent, and electrowinning, in which metallic ions are deposited onto an electrode by an electric current passed through the solution.

Extraction is often followed by refining, in which the level of impurities is brought lower or controlled by pyrometallurgical, electrolytic, or chemical means. Pyrometallurgical refining usually consists of the oxidizing of impurities in a high-temperature liquid bath. Electrolysis is the dissolving of metal from one electrode of an electrolytic cell and its deposition in a purer form onto the other electrode. Chemical refining involves either the condensation of metal from a vapour or the selective precipitation of metal from an aqueous solution.

The processes to be used in extraction and refining are selected to fit into an overall pattern, with the product from the first process becoming the feed material of the second process, and so on. It is quite common for hydrometallurgical, pyrometallurgical, and electrolytic processes to be used one after another in the treatment of a single metal. The choices depend on several conditions. One of these is that certain types of metallic compounds lend themselves to easiest extraction by certain methods; for example, oxides and sulfates are readily dissolved in leach solutions, while sulfides are only slightly soluble. Another condition is the degree of purity, which can vary from one type of extraction to another. Zinc production illustrates this, in that zinc metal produced by pyrometallurgical retort or blast-furnace operations is 98 percent pure, with traces of lead, iron, and cadmium. This is adequate for galvanizing, but the preferred purity for die-casting (99.99 percent) must be obtained hydrometallurgically, from the electrolysis of a zinc sulfate solution. Also to be considered in choosing a processing method is the recovery of particular impurities that may have value themselves as by-products. One example is copper refining: the pyrometallurgical refining of blister copper removes many impurities, but it does not recover or remove silver or gold; the precious metals are recovered, however, by subsequent electrolytic refining.

Pyrometallurgy

Two of the most common pyrometallurgical processes, in both extraction and refining, are oxidation and reduction. In oxidation, metals having a great affinity for oxygen selectively combine with it to form metallic oxides; these can be treated further in order to obtain a pure metal or can be separated and discarded as a waste product. Reduction can be viewed as the reverse of oxidation. In this process, a metallic oxide compound is fed into a furnace along with a reducing agent such as carbon. The metal releases its combined oxygen, which recombines with the carbon to form a new carbonaceous oxide and leaves the metal in an uncombined form.

Oxidation and reduction reactions are either exothermic (energy-releasing) or endothermic (energy-absorbing). One example of an exothermic reaction is the oxidation of iron sulfide (FeS) to form iron oxide (FeO) and sulfur dioxide (SO_2) gas:

$$2\text{FeS} + 3\text{O}_2 \rightarrow 2\text{Fe}O + 2\text{SO}_2$$

This process gives off large quantities of heat beyond that required to initiate the reaction. One endothermic reaction is the smelting reduction of zinc oxide (ZnO) by carbon monoxide (CO) to yield zinc (Zn) metal and carbon dioxide (CO_2):

$$ZnO + CO \rightarrow Zn + CO_2$$

For this reaction to proceed at a reasonable rate, external heat must be supplied to maintain the temperature at 1,300 to 1,350 °C (2,375 to 2,450 °F).

Roasting

As stated above, for those instances in which a metal-bearing compound is not in a chemical form that permits the metal to be easily and economically removed, it is necessary first to change it into some other compound. The preliminary treatment that is commonly used to do this is roasting.

Processes

There are several different types of roast, each one intended to produce a specific reaction and to yield a roasted product (or calcine) suitable for the particular processing operation to follow. The roasting procedures are:

1. Oxidizing roasts, which remove all or part of the sulfur from sulfide metal compounds, replacing the sulfides with oxides. (The sulfur removed goes off as sulfur dioxide gas.) Oxidizing roasts are exothermic.
2. Sulfatizing roasts, which convert certain metals from sulfides to sulfates. Sulfatizing roasts are exothermic.
3. Reducing roasts, which lower the oxide state or even completely reduce an oxide to a metal? Reducing roasts are exothermic.
4. Chloridizing roasts, or chlorination, which change metallic oxides to chlorides by heating with a chlorine source such as chlorine gas, hydrochloric acid gas, ammonium chloride, or sodium chloride. These reactions are exothermic.
5. Volatilizing roasts, which eliminate easily volatilized oxides by converting them to gases.
6. Calcination, in which solid material is heated to drive off either carbon dioxide or chemically combined water. Calcination is an endothermic reaction.

Roasters

Each of the above processes can be carried out in specialized roasters. The types most commonly in use are fluidized-bed, multiple-hearth, flash, chlorinator, rotary kiln, and sintering machine (or blast roaster).

Fluidized-bed roasters have found wide acceptance because of their high capacity and efficiency. They can be used for oxidizing, sulfatizing, and volatilizing roasts. The roaster is a refractory-lined, upright cylindrical steel shell with a grate bottom through which air is blown in sufficient volume to keep fine, solid feed particles in suspension and give excellent gas-solid contact. The ore feed can be introduced dry or as a water suspension through a downpipe into the turbulent layer zone of the roaster. Discharge of the roasted calcines is through a side overflow pipe.

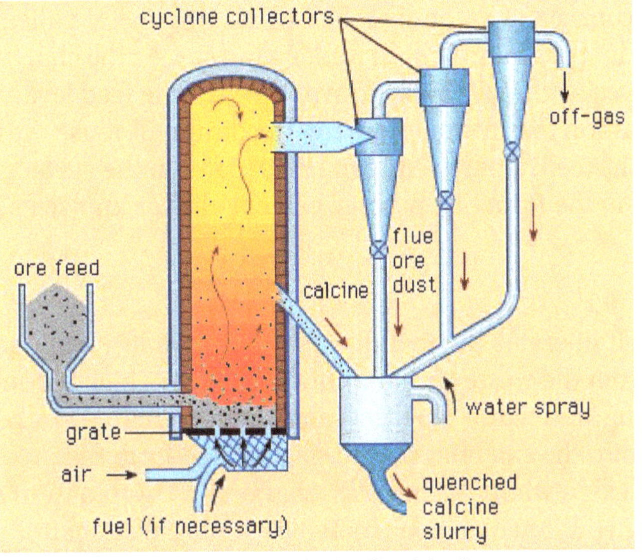

Schematic diagram of a fluidized-bed roaster.

Multiple-hearth roasters also have found wide acceptance in that they can be used for oxidizing, sulfatizing, chloridizing, volatilizing, reducing, and calcining processes. The roaster is a refractory-lined, vertical cylindrical steel shell in which are placed a number of superimposed refractory hearths. A slowly rotating central shaft turns rabble arms on each hearth both to stir the roasting material and to push it into drop holes leading to the hearth below. Feed material is fed to the top hearth, and, as it follows a zig-zag path across the hearths and downward, it meets the rising gas stream that effects the roasting. The calcines are discharged from the bottom hearth.

Flash roasters are used only for oxidizing roasts and are, in effect, multiple-hearth roasters with the central hearths removed. This design came with the realization that much of the oxidizing takes place as the particles are actually dropping from hearth to hearth.

Chlorinators are used for roasting oxides to chlorides. They are tall, circular steel shells lined with refractory brick to prevent chlorine attack on the steel. The top of each chlorinator has a sealed hopper for periodic feed charging, and gaseous or liquid chlorine is added at the bottom of the unit. Heat is supplied by electrical resistance through the shell wall and by any exothermic reaction that may occur. The product depends on the chloridizing reaction taking place, with magnesium dichloride, for example, forming as a watery liquid and titanium tetrachloride coming off as a gas.

Calcination of carbonates to oxides is done in a horizontal rotary kiln, which is a mild-steel circular shell lined with refractory material and having a length 10 to 12 times the diameter. Sloping slightly downward from feed to discharge ends, the kiln slowly rotates while fuel-fired burners located inside the kiln provide the required heat.

A sintering machine, or blast roaster, can conduct oxidizing or reducing roasts and then agglomerate the roasted calcines, or it can be used for agglomeration alone. (Agglomeration is the fusing of fine feed material into larger chunks that can be fed into a blast furnace or retort, eliminating the problem of losing the fine feed in the hot air blast.) The oxidizing or reducing reaction is exothermic, but in order for agglomeration alone to be conducted, a fuel such as fine coke must be mixed with the charge.

The sintering machine consists of an endless belt of moving metal pallets with grate bottoms on which a fine feed charge is spread and passed under a burner. As the charge ignites, the pallet passes over a suction wind box, so that air being drawn through the feed layer causes combustion (*i.e.*, oxidation) of sulfur or carbon to continue from top to bottom. Because the temperature is high and there is no agitation of the feed, a partial fusion takes place on the surface of the particles, leaving them adhering together in the form of a porous, cellular clinker known as sinter.

Smelting

Smelting is a process that liberates the metallic element from its compound as an impure molten metal and separates it from the waste rock part of the charge, which becomes a molten slag. There are two types of smelting, reduction smelting and matte smelting. In reduction smelting, both the metallic charge fed into the smelter and the slag formed from the process are oxides; in matte smelting, the slag is an oxide while the metallic charge is a combination of metallic sulfides that melt and recombine to give a homogeneous metallic sulfide called matte.

Reduction Smelting

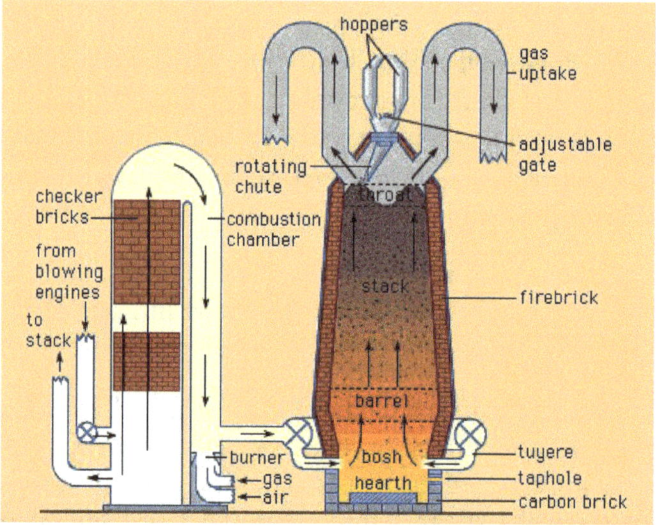

Schematic diagram of modern blast furnace (right) and hot-blast stove (left).

Many types of furnace are used for reduction smelting. The blast furnace is universally used in the reduction of such compounds as iron oxide, zinc oxide, and lead oxide, though there are great differences between the furnaces designs used in each case. Iron, found naturally in the oxide ores hematite and magnetite, is smelted in a tall, circular, sealed blast furnace. A sintered or pelletized feed consisting of coke (for fuel), limestone (as a flux for slag making), and iron oxide is charged into the top of the furnace through a double bell or rotating chute, and heated air is blown in through nozzles, or tuyeres, close to the furnace bottom. In the ensuing combustion reaction, oxygen in the air combines with carbon in the coke, generating enough heat to melt the furnace charge and forming carbon monoxide, which, in turn, reduces the iron oxide to metallic iron. The furnace is sealed to prevent the escape of carbon monoxide gas, which is recovered and burned as fuel to heat the tuyere air. In the hearth at the furnace bottom, molten slag and iron collect in two layers, the lighter slag on top. Both are periodically drawn off, with the slag being discarded and the iron going on to be refined into steel.

The zinc blast furnace also is a sealed furnace, with a charge of sintered zinc oxide and preheated coke added through a sealed charging bell. The furnace is rectangular, with a shorter shaft than the iron blast furnace. A blast of hot air through the tuyeres provides oxygen to burn the coke for heat and to supply carbon monoxide reducing gas. The reduced zinc passes out of the furnace as vapour, and this is drawn off to a spray bath of molten lead and condensed to liquid zinc metal. Slag and any lead present in the charge are tapped as liquids from the furnace hearth.

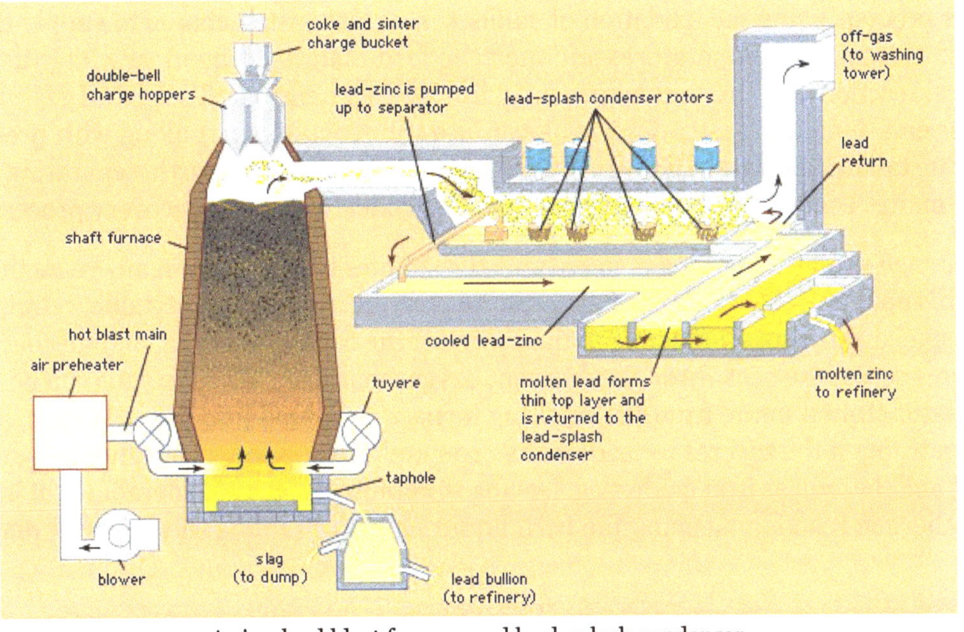

A zinc-lead blast furnace and lead-splash condenser.

The lead blast furnace is similar in size and shape to the zinc blast furnace, but it is not a sealed furnace, and it does not use preheated tuyere air. A charge of lead oxide sinter, coke, and flux is poured into the open top of the furnace, and the strong reducing atmosphere in the furnace shaft reduces the oxide to metal. Liquid lead and slag collect in two layers in the furnace hearth, with the lead in the bottom layer and slag above.

Two newer processes for the direct reduction of unroasted lead sulfide concentrate are the QSL (Queneau-Schuhmann-Lurgi) and the KIVCET (a Russian acronym for "flash-cyclone-oxygen-electric smelting"). In the QSL reactor a submerged injection of shielded oxygen oxidizes lead sulfide to lead metal, while the KIVCET is a type of flash-smelting furnace in which fine, dried lead sulfide concentrate combines with oxygen in a shaft to give lead metal.

Matte Smelting

The primary purpose of matte smelting is to melt and recombine the charge into a homogeneous matte of metallic copper, nickel, cobalt, and iron sulfides and to give an iron and silicon oxide slag. It is done in many types of furnace on either roasted or unroasted sulfide feed material.

The reverberatory furnace is essentially a rectangular refractory-brick box equipped with end-wall burners to provide heat for melting. The furnace is relatively quiet, and it does not blow out much fine feed (which is added through roof ports) with the exhaust gases. The matte is tapped periodically from a centre taphole, while the slag runs off continuously at the furnace flue end. Oxygen lances inserted through the roof, or oxygen added through the burners, can increase smelting capacity considerably.

Electric furnaces are similar to reverberatory furnaces except for the method of heating—in this case a row of electrodes projecting through the roof into the slag layer on the furnace hearth and heating by resistance.

Flash smelting is a relatively recent development that has found worldwide acceptance. It is an autogenous process, using the oxidation of sulfides in an unroasted charge to supply the heat required to reach reaction temperatures and melt the feed material. The most widely used furnace has a vertical reaction shaft at one end of a long, low settling hearth and a vertical gas-uptake shaft at the other end. Fine, unroasted feed is blown into the reaction shaft along with preheated air; these react instantaneously, and liquid droplets fall onto the settling hearth, separating into layers of slag and matte. The off-gas, high in sulfur dioxide, is ideal for sulfur-recovery processes.

The second stage of matte smelting is converting the sulfides to metal. For many years the standard vessel for this operation has been the Peirce-Smith converter. This is a rotatable, refractory-lined, horizontal steel drum with an opening at the centre of the top for charging and discharging and a row of tuyeres across the back through which air, oxygen-enriched air, or oxygen can be blown into the liquid bath. Molten matte from the smelting furnace is poured into the converter, after which gas is blown through the tuyeres to oxidize first iron and then sulfur. The sulfur goes off as sulfur dioxide gas and the iron as iron oxide slag, leaving semipure metal. Considerable heat is generated by this exothermic reaction, keeping the bath liquid and maintaining the required reaction temperature.

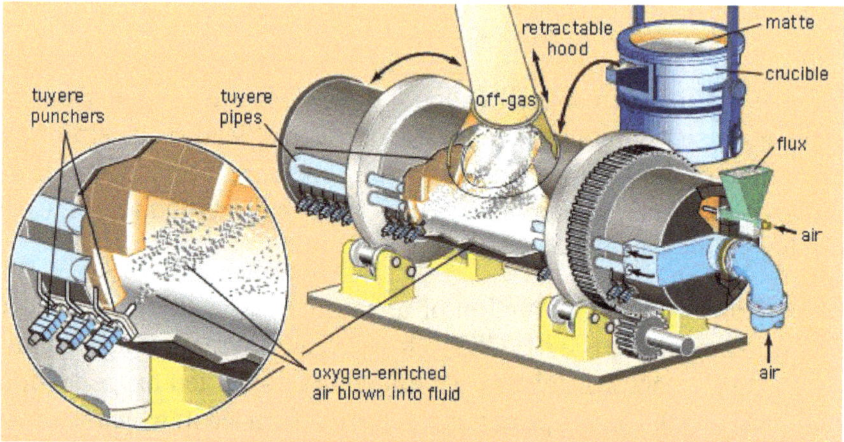

A side-blown copper-nickel matte converter.

More recent processes take advantage of exothermic heat evolution to accomplish both the smelting of unroasted sulfides and the conversion of matte in one combined operation. These are the Noranda, TBRC (top-blown rotary converter), and Mitsubishi processes. The Noranda reactor is a horizontal cylindrical furnace with a depression in the centre where the metal collects and a raised hearth at one end where the slag is run off. Pelletized unroasted sulfide concentrate is poured into the molten bath at one end, where tuyeres inject an air-oxygen mixture. This causes an intense mixing action that aids the melting, smelting, and oxidation steps, which follow one another in sequence, by taking advantage of the exothermic heat. The TBRC also is cylindrical in shape but is inclined at 17° to the horizontal, has an open mouth at the high end for charging and pouring, and revolves at 5 to 40 rotations per minute. A lance inserted through the mouth can give any combination of oxygen, air, or natural gas to impinge on the molten bath and create the conditions required for smelting and oxidizing. The combination of surface blowing and bath rotation improves the performance of the converter. The Mitsubishi process is a continuous smelting-converting operation that uses three stationary furnaces in series. The first furnace is for smelting, with oxygen lances and a fuel-fired burner inserted through the roof. Slag and matte flow from here to a slag-cleaning furnace (heated by electric arc), and high-grade matte flows from this to the converting furnace, where oxygen-enriched air is blown into the bath through roof lances. Exothermic heat produced here is sufficient to keep the bath up to reaction temperature.

Electrolytic Smelting

Smelting is also carried out by the electrolytic dissociation, at high temperatures, of a liquid metallic chloride compound (as is done with magnesium) or of a metallic oxide powder dissolved in molten electrolyte (as is done with aluminum). In each case, electric current is passed through the bath to dissociate the metallic compound; the metal released collects at the cathode, while a gas is given off at the anode.

The magnesium smelting cell consists of a steel pot that serves as the cathode; two rows of graphite electrodes are inserted through a refractory cover as anodes. The electrolyte is a mixture of chlorides, with magnesium chloride making up 20 percent, and the cell is maintained at 700 °C (1,300 °F). The passage of current breaks down the magnesium chloride into chlorine gas and magnesium metal, which go to the anode and cathode, respectively.

In the Hall-Héroult smelting process, a nearly pure aluminum oxide compound called alumina is dissolved at 950 °C (1,750 °F) in a molten electrolyte composed of aluminum, sodium, and fluorine; this is electrolyzed to give aluminum metal at the cathode and oxygen gas at the anode. The smelting cell is a carbon-lined steel box, which acts as the cathode, and a row of graphite electrodes inserted into the bath serves as anodes.

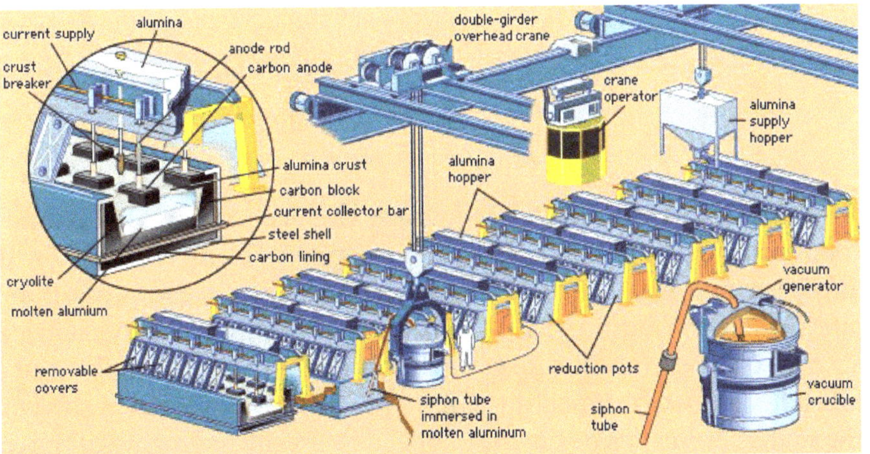

Part of a modern potline based on the electrolytic Hall-Héroult smelting process.

Refining

Refining is the final procedure for removing (and often recovering as by-products) the last small amounts of impurities left after the major extraction steps have been completed. It leaves the major metallic element in a practically pure state for commercial application. The procedure is accomplished in three ways: refining by fire, by electrolytic, or by chemical methods.

Fire Refining

Iron, copper, and lead are fire-refined by selective oxidation. In this process, oxygen or air is added to the impure liquid metal; the impurities oxidize before the metal and are removed as an oxide slag or a volatile oxide gas.

The basic oxygen furnace (BOF) is a vessel used to convert pig iron, of about 94 percent iron and 6 percent combined impurities such as carbon, manganese, and silicon, into steel with as little as 1 percent combined impurities. The BOF is a large pear-shaped unit that can be tilted to charge and pour. Molten blast-furnace iron and steel scrap are charged into the furnace; then it is turned to an upright position and a lance inserted to blow high-tonnage oxygen gas into the bath. Oxidation reactions occur rapidly, with silicon and manganese oxidizing first and combining to form an oxide slag, then carbon oxidizing to carbon monoxide gas and burning to carbon dioxide as it leaves the furnace mouth. These reactions are strongly exothermic and keep the vessel up to its reaction temperature without any external heat or fuel being added.

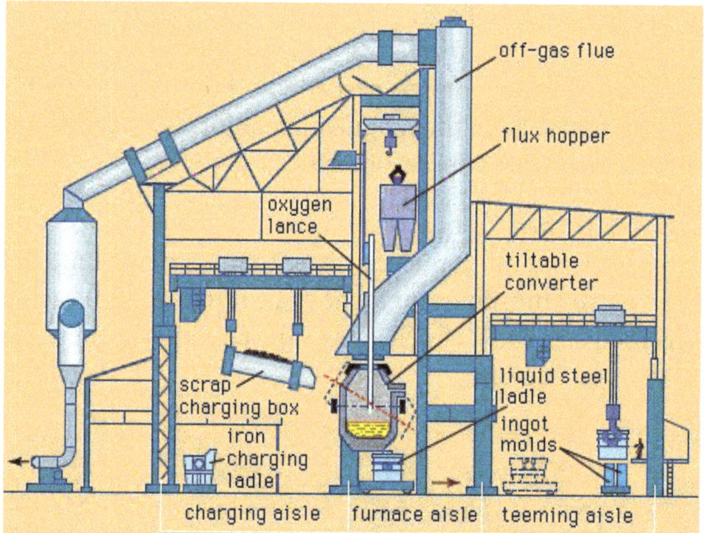

A basic oxygen furnace shop.

Converter-produced blister copper and blast-furnace lead also are treated by fire refining, with both processes depending on the weaker affinity for oxygen of the metals than the impurities they contain. Molten copper in a small reverberatory-type furnace has compressed air blown into it through steel pipes below the surface. This oxidizes zinc, tin, iron, lead, arsenic, antimony, and sulfur; the sulfur goes off as sulfur dioxide gas, while the other impurities form an oxide slag that is skimmed off. Lead is refined in much the same way, with compressed air blown into a molten lead bath and the major impurities of tin, antimony, and arsenic oxidizing in that order, rising to the surface as skims and being scraped off.

Other fire-refining operations use fractional distillation. By this method, zinc metal of 98 percent purity can be upgraded to 99.995 percent purity. The main impurities in blast-furnace zinc are lead and cadmium, with lead boiling at 1,744 °C (3,171 °F), zinc at 907 °C (1,665 °F), and cadmium at 765 °C (1,409 °F). In the first stage zinc and cadmium are boiled off, leaving liquid lead, and in the second stage cadmium is boiled off to leave special high-purity zinc metal.

Electrolytic Refining

This method gives the highest-purity metal product as well as the best recovery of valuable impurities. It is used for copper, nickel, lead, gold, and silver. The metal to be refined is cast into a slab, which becomes the anode of an electrolytic cell; another sheet of metal is the cathode. Both electrodes are immersed in an aqueous electrolyte capable of conducting an electric current. As a direct current is impressed on the cell, metal ions dissolve from the anode and deposit at the cathode. The insoluble sludge left in the cell is treated to recover any valuable by-product metals.

Chemical Refining

An example of chemical refining is the nickel carbonyl process, in which impure nickel metal is selectively reacted with carbon monoxide gas to form nickel carbonyl gas. This gas is then decomposed to give high-purity nickel metal.

Hydrometallurgy

Hydrometallurgy is concerned with the selective leaching of metallic compounds to form a solution from which the metals can be precipitated and recovered. Leaching processes are used when it is the simplest method or when the ore is of too low a grade for more expensive extractive procedures.

Conversion

Because not all ores and concentrates are found naturally in a form that is satisfactory for leaching, they must often be subjected to preliminary operations. For example, sulfide ores, which are relatively insoluble in sulfuric acid, can be converted to quite soluble forms by oxidizing or sulfatizing roasts. On the other hand, oxide ores and concentrates can be given a controlled reducing roast in order to produce a calcine containing a reduced metal that will dissolve easily in the leaching solution.

A second popular treatment for converting sulfides is pressure oxidation, in which the sulfides are oxidized to a porous structure that provides good access for the leaching solution. This treatment was developed for the recovery of gold from sulfide ores, which are not suitable for cyanide leaching without first being oxidized. A finely ground concentrate slurry is preheated to 175 °C (350 °F) and pumped into a four- or five-compartment autoclave, each compartment containing an agitator. Gaseous oxygen is added to each compartment, and retention time in the autoclave is two hours in order to achieve the desired oxidation.

Leaching

Oxides are leached with a sulfuric acid or sodium carbonate solvent, while sulfates can be leached with water or sulfuric acid. Ammonium hydroxide is used for native ores, carbonates, and sulfides, and sodium hydroxide is used for oxides. Cyanide solutions are a solvent for the precious metals, while a sodium chloride solution dissolves some chlorides. In all cases the leach solvent should be cheap and available, strong, and preferably selective for the values present.

Leaching is carried out by two main methods: simple leaching at ambient temperature and atmospheric pressure; and pressure leaching, in which pressure and temperature are increased in order to accelerate the operation. The method chosen depends on the grade of the feed material, with richer feed accommodating a costlier, more extensive treatment.

Leaching in-place, or in situ leaching, is practiced on ores that are too far underground and of too low a grade for surface treatment. A leach solution is circulated down through a fractured ore body to dissolve the values and is then pumped to the surface, where the values are precipitated.

Heap leaching is done on ores of semi-low grade—that is, high enough to be brought to the surface for treatment. This method is increasing in popularity as larger tonnages of semi-low-grade ore are mined. The ore is piled in heaps on pads and sprayed with bleach solution, which trickles down through the heaps while dissolving the values. The pregnant solution is drained away and taken to precipitation tanks.

Higher-grade ores are treated by tank leaching, which is carried out in two ways. One method is of very large scale, with several thousand tons of ore treated at a time in large concrete tanks with a circulating solution. In the second method, small amounts of finely ground high-grade ore are agitated in tanks by air or by mechanical impellers. Both solutions pass to precipitation after leaching is completed.

Pressure leaching shortens the treatment time by improving the solubility of solids that dissolve only very slowly at atmospheric pressure. For this process autoclaves are used, in both vertical and horizontal styles. After leaching, the pregnant solution is separated from the insoluble residue and sent to precipitation.

Recovery

Pregnant solution from leaching operations is treated in a variety of ways to precipitate the dissolved metal values and recover them in solid form. These include electrolytic deposition, transfer of metal ions, chemical precipitation, solvent extraction in combination with electrolytic and chemical methods, and carbon adsorption combined with electrolytic treatment.

Electrolytic deposition, also called electro winning, gives a pure product and is a preferred method. However, it is expensive, owing to the cost of electricity, and must have a solution of high metal content. Insoluble anodes, and cathodes made of either a strippable inert material or a thin sheet of the deposited metal, are inserted into a tank containing leach solution. As current is passed, the solution dissociates, and metal ions deposit at the cathode. This common method is used for copper, zinc, nickel, and cobalt.

Solvent extraction combined with electrolytic deposition takes dilute, low-value metal solutions and concentrates them into small volumes and high metal contents, rendering them satisfactory for electrolytic treatment. Low-grade copper ores are processed in this manner. First, a large volume of a low-value copper leach solution (2.5 grams per litre, or 0.33 ounces per gallon) is contacted with a small volume of water-immiscible organic solvent in kerosene. The metal values pass from the leach solution into the extraction solution, the two phases are separated, and the extraction solution goes on to the stripping circuit. Here another fluid is added that has a still greater affinity for the metal values, picking them out of the extraction solution. The two solutions are separated, with the small volume of stripping solution having a metal content high enough (50 grams per litre, or 6.6 ounces per gallon) to be suitable for electrolytic precipitation.

An adsorption circuit is used to strip pregnant solutions of gold cyanide with activated carbon. The carbon is in turn stripped of the metal by a solution, which then goes to an electrolytic cell where the gold content is deposited at the cathode.

Chemical precipitation can be accomplished in a number of ways. In one method, a displacement reaction takes place in which a more active metal replaces a less active metal in solution. For example, in copper cementation iron replaces copper ions in solution, solid particles of copper precipitating while iron goes into solution. This is an inexpensive method commonly applied to weak, dilute leach solutions. Another displacement reaction uses gas, with hydrogen sulfide, for example, added to a solution containing nickel sulfate and precipitating nickel sulfide. Finally, changing the acidity of a solution is a common method of precipitation. Yellow cake, a common name for sodium diuranate, is precipitated from a concentrated uranium leach solution by adding sodium hydroxide to raise the pH to 7.

Physical Metallurgy

Physical metallurgy is the science of making useful products out of metals. Metal parts can be made in a variety of ways, depending on the shape, properties, and cost desired in the finished product. The desired properties may be electrical, mechanical, magnetic, or chemical in nature; all of them can be enhanced by alloying and heat treatment. The cost of a finished part is often determined more by its ease of manufacture than by the cost of the material. This has led to a wide variety of ways to form metals and to an active competition among different forming methods, as well as among different materials. Large parts may be made by casting. Thin products such as automobile fenders are made by forming metal sheets, while small parts are often made by powder metallurgy (pressing powder into a die and sintering it). Usually a metal part has the same properties throughout. However, if only the surface needs to be hard or corrosion-resistant, the desired performance can be obtained through a treatment that changes only the composition and strength of the surface.

Structures and Properties of Metals

Metallic Crystal Structures

Metals are used in engineering structures (*e.g.,* automobiles, bridges, pressure vessels) because, in contrast to glass or ceramic, they can undergo appreciable plastic deformation before breaking. This plasticity stems from the simplicity of the arrangement of atoms in the crystals making up a piece of metal and the non-directional nature of the bond between the atoms. Atoms can be arranged in many different ways in crystalline solids, but in metals the packing is in one of three simple forms. In the most ductile metals, atoms are arranged in a close-packed manner. If atoms were visualized as identical spheres and if these spheres were packed into planes in the closest possible manner, there would be two ways to stack close-packed planes one above another. One would lead to a crystal with hexagonal symmetry (called hexagonal close-packed, or hcp); the other would lead to a crystal with cubic symmetry that could also be visualized as an assembly of cubes with atoms at the corners and at the centre of each face (known as face-centred cubic, or fcc). Examples of metals with the hcp type of structure are magnesium, cadmium, zinc, and alpha titanium. Metals with the fcc structure include aluminum, copper, nickel, gamma iron, gold, and silver.

Common metallic crystal structures

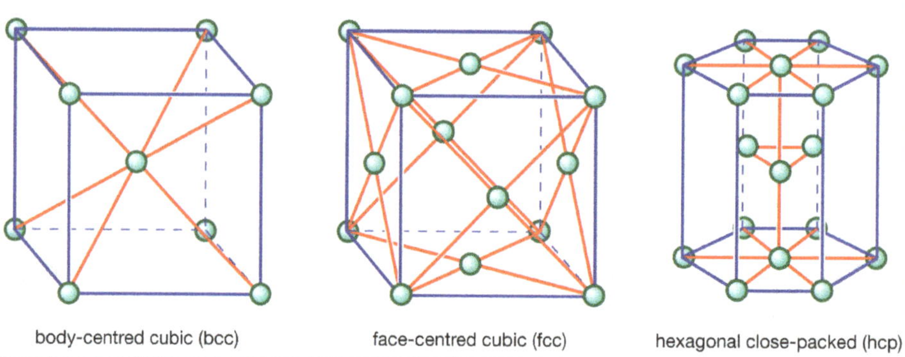

The third common crystal structure in metals can be visualized as an assembly of cubes with atoms at the corners and an atom in the centre of each cube; this is known as body-centred cubic, or bcc. Examples of metals with the bcc structure are alpha iron, tungsten, chromium, and beta titanium.

Some metals, such as titanium and iron, exhibit different crystal structures at different temperatures. The lowest-temperature structure is labeled alpha (α), and higher-temperature structures beta (β), gamma (γ), and delta (δ). This allotropy, or transformation from one structure to another with changing temperature, leads to the marked changes in properties that can come from heat treatment.

When a metal undergoes a phase change from liquid to solid or from one crystal structure to another, the transformation begins with the nucleation and growth of many small crystals of the new phase. All these crystals, or grains, have the same structure but different orientations, so that, when they finally grow together, boundaries form between the grains. These boundaries play an important role in determining the properties of a piece of metal. At room temperature they strengthen the metal without reducing its ductility, but at high temperatures they often weaken the structure and lead to early failure. They can be the site of localized corrosion, which also leads to failure.

Mechanical Properties

When a metal rod is lightly loaded, the strain (measured by the change in length divided by the original length) is proportional to the stress (the load per unit of cross-sectional area). This means that, with each increase in load, there is a proportional increase in the rod's length, and, when the load is removed, the rod shrinks to its original size. The strain here is said to be elastic, and the ratio of stress to strain is called the elastic modulus. If the load is increased further, however, a point called the yield stress will be reached and exceeded. Strain will now increase faster than stress, and, when the sample is unloaded, a residual plastic strain (or elongation) will remain. The elastic strain at the yield stress is typically 0.1 to 1 percent, whereas, with the sample pulled to rupture, the plastic strain is typically 20 to 40 percent for an alloy (it may exceed 100 percent in some cases).

The most important mechanical properties of a metal are its yield stress, its ductility (measured by the elongation to fracture), and its toughness (measured by the energy absorbed in tearing the metal). The yield stress of a metal is determined by the resistance to slipping of one plane of atoms over another. Various barriers to slip can be produced by heat treatment and alloying; examples of such barriers are grain boundaries, fine precipitates, distortion introduced by cold working the metal, and alloying elements dissolved in the metal.

When a metal is made very strong through one or more of these methods, it may suddenly fracture under a load instead of yielding. This is particularly true when the metal contains notches or cracks that locally raise the stress and localize the yielding. The property of interest then becomes the fracture toughness, measured by the energy required to extend an existing crack in a piece of metal. In almost all cases, the fracture toughness of an alloy can be improved only by reducing its yield strength. The only exception to this is a smaller grain size, which increases both toughness and strength.

Electrical Properties

The electrical conductivity of a metal (or its reciprocal, electrical resistivity) is determined by the ease of movement of electrons past the atoms under the influence of an electric field. This movement is particularly easy in copper, silver, gold, and aluminum—all of which are well-known conductors of electricity. The conductivity of a given metal is decreased by phenomena that deflect, or scatter, the moving electrons. These can be anything that destroys the local perfection of the atomic arrangement—for example, impurity atoms, grain boundaries, or the random oscillation of atoms induced by thermal energy. This last example explains why the conductivity of a metal increases substantially with falling temperature: in a pure metal at room temperature, most resistance to the motion of free electrons comes from the thermal vibration of the atoms; if the temperature is reduced to almost absolute zero, where thermal motion essentially stops, conductivity can increase several thousandfold.

Magnetic Properties

When an electric current is passed through a coil of metal wire, a magnetic field is developed around the coil. When a piece of copper is placed inside the coil, this field increases by less than 1 percent, but, when a piece of iron, cobalt, or nickel is placed inside the coil, the external field can increase 10,000 times. This strong magnetic property is known as ferromagnetism, and the three metals listed above are the most prominent ferromagnetic metals. When the piece of ferromagnetic metal is removed from the coil, it retains some of this magnetism (that is, it is magnetized). If the metal is hard, as in a hardened piece of steel, the loss, or reversal, of magnetization will be slow, and the sample will be useful as a permanent magnet. If the metal is soft, it will quickly lose its magnetism; this will make it useful in electrical transformers, where rapid reversal of magnetization is essential.

In many types of solids, the atoms possess a permanent magnetic moment (they act like small bar magnets). In most solids, the direction of these moments is arranged at random. What is exceptional about ferromagnetic solids is that the interatomic forces cause the moments of neighbouring atoms spontaneously to align in the same direction. If the moments of all of the atoms in a single sample lined up in the same direction, the sample would be an exceptionally strong magnet with exceptionally high energy. That energy would be reduced if the sample broke up into domains, with all atomic moments in each domain being aligned but the direction of magnetization in adjacent domains being in opposite directions and thus tending to cancel one another. This is what happens when a ferromagnetic metal is magnetized: all domains do not take on the same orientation, but domains of one orientation grow at the expense of others. The alignment of atomic magnetic moments within a domain is weakened by thermally induced oscillations, and ferromagnetism is finally lost above the Curie point, which is 770 °C (1,420 °F) for iron and 358 °C (676 °F) for nickel.

Chemical Properties

Almost any metal will oxidize in air, the only exception being gold. At room temperature a clean metal surface will oxidize very little, since a thin oxide film forms and protects the metal from further oxidation. At elevated temperatures, though, oxidation is faster, and the film is less protective. Many chemicals accelerate this corrosion process (that is, the conversion of a metal to an oxide in air or to a hydroxide in the presence of water).

A special property of metal surfaces is their ability to catalyze chemical reactions. For example, in the exhaust system of most automobiles, combustion gases pass over a dispersion of very fine platinum particles. The surfaces of these particles greatly accelerate the oxidation of carbon monoxide and hydrocarbons to carbon dioxide and water, thus reducing the toxicity of the exhaust gases.

Alloying

Almost all metals are used as alloys—that is, mixtures of several elements—because these have properties superior to pure metals. Alloying is done for many reasons, typically to increase strength, increase corrosion resistance, or reduce costs.

Processes

In most cases, alloys are mixed from commercially pure elements. Mixing is relatively easy in the liquid state but slow and difficult in the solid state, so that most alloys are made by melting the base metal—for instance, iron, aluminum, or copper—and then adding the alloying agents. Care must be taken to avoid contamination, and in fact purification is often carried out at the same time, since this is also done more easily in the liquid state. Examples can be found in steelmaking, including the desulfurizing of liquid blast-furnace iron in a ladle, the decarburization of the iron during its conversion to steel, the removal of oxygen from the liquid steel in a vacuum degasser, and finally the addition of tiny amounts of alloying agents to bring the steel to the desired composition.

The largest tonnages of alloys are melted in air, with the slag being used to protect the metal from oxidation. However, a large and increasing amount is melted and poured entirely in a vacuum chamber. This allows close control of the composition and minimizes oxidation. Most of the alloying elements needed are placed in the initial charge, and melting is done with electricity, either by induction heating or by arc melting. Induction melting is conducted in a crucible, while in arc melting the melted droplets drip from the arc onto a water-cooled pedestal and are immediately solidified.

Sometimes an inhomogeneous, composite structure is desired, as in cemented tungsten carbide cutting tools. In such cases, the alloy is not melted but is made by powder metallurgical techniques.

Metallurgy

Increasing Strength

The most common reason for alloying is to increase the strength of a metal. This requires that barriers to slip be distributed uniformly throughout the crystalline grains. On the finest scale, this is done by dissolving alloying agents in the metal matrix (a procedure known as solid solution hardening). The atoms of the alloying metals may substitute for matrix atoms on regular sites (in which case they are known as substitutional elements), or, if they are appreciably smaller than the matrix atoms, they may take up places between regular sites (where they are called interstitial elements).

The next coarser type of barrier to slip is a fine, solute-rich precipitate with dimensions of only tens or hundreds of atomic diameters. These particles are formed by heat treatment. The metal is heated to a temperature at which the solute-rich phase dissolves (*e.g.*, 5 percent copper in aluminum at 540 °C [1,000 °F]), and then it is rapidly cooled to avoid precipitation. The next step is to form a fine precipitate throughout the sample by aging at an elevated temperature that is well below the temperature used for the initial dissolution.

In metals that undergo transformations from one crystal structure to another on heating (*e.g.*, iron or titanium), the difference in solute solubility between the high- and low-temperature phases is often utilized. For example, in the low-alloy steels used for tools and gears, carbon forms the hardening precipitate. Carbon is much more soluble in the high-temperature fcc phase (gamma iron, also called austenite) than in the low-temperature bcc phase (alpha iron, or ferrite). The other alloying elements added (*e.g.*, chromium, nickel, and molybdenum) retard the transformation of austenite on cooling, so that the fcc-to-bcc transformation occurs at a low temperature by a sudden, shear transformation; this allows no time for carbon precipitation and makes the steel harder. A final reheating tends to coarsen the precipitate and thereby increase ductility; this is commonly called tempering.

An array of barriers on the same scale as precipitation hardening can be created by plastically deforming the metal at room temperature. This is often done in a cold-working operation such as rolling, forging, or drawing. The deformation occurs through the generation and motion of line defects, called dislocations, on slip planes spaced only a few hundred atom diameters apart. When slip occurs on different planes, the intersecting dislocations form tangles that inhibit further slip on those planes. Such strain hardening can double or triple the yield stress of a metal.

Increasing Corrosion Resistance

Alloys can have much better high-temperature oxidation resistance than pure metals. The alloying elements most commonly used for this purpose are chromium and aluminum, both of which form an adherent film of stable oxide on the surface that protects the metal from further oxidation. Eleven percent or more chromium is added to iron to create a stainless steel, while 10 to 15 percent chromium and 3 to 5 percent aluminum are commonly added to the nickel- or cobalt-based superalloys used in the highest-temperature components of jet engines.

Inhibiting the corrosion of alloys in water is more varied and complex than inhibiting high-temperature oxidation. Nevertheless, one of the most common techniques is to add alloying elements that inhibit the corrosion.

Reducing Costs

Gold and silver used in jewelry and coins are alloyed with other metals to increase strength and reduce cost. Sterling silver contains 7.5 percent base metal, commonly copper. The fraction of gold in gold jewelry is designated in karats, with 24-karat being pure gold and 18-karat being 75 percent gold by weight. In coins, alloys with the look and density of silver are commonly substituted for silver; for instance, all U.S. coins that appear to be made of silver actually have a surface layer of 75 percent copper and 25 percent nickel.

Lowering Melting Points

Alloying can also be done to lower the melting point of a metal. For example, adding lead to tin lowers the melting point of the tin-rich alloy, and adding tin to lead lowers the melting point of the lead-rich alloy. A 62-percent-tin 38-percent-lead alloy, which is called the eutectic composition, has the lowest melting point of all, much lower than that of either metal. Eutectic lead-tin alloys are used for soldering.

Casting

Casting consists of pouring molten metal into a mold, where it solidifies into the shape of the mold. The process was well established in the Bronze Age (beginning c. 3000 BC), when it was used to form most of the bronze pieces now found in museums. It is particularly valuable for the economical production of complex shapes, ranging from mass-produced parts for automobiles to one-of-a-kind production of statues, jewelry, or massive machinery.

Processes

Casting processes differ in how the mold is made and in how the metal is forced into the mold. For metals with a high melting temperature, stable refractory material must be used to avoid reaction between the metal and the mold. Most steel and iron castings, for example, are poured into silica sand, though some parts are cast into coated metal molds. For metals of lower melting point, such as aluminum or zinc, molds can be made of another metal or of sand, depending on how many parts are to be produced and other considerations. Gravity is most frequently employed to fill the mold, but some processes use centrifugal force or pressure injection.

Sand-casting

Sand-casting is widely used for making cast-iron and steel parts of medium to large size in which surface smoothness and dimensional precision are not of primary importance.

The first step in any casting operation is to form a mold that has the shape of the part to be made. In many processes, a pattern of the part is made of some material such as wood, metal, wax, or polystyrene and refractory molding material is formed around this. For example, in greensand-casting, sand combined with a binder such as water and clay is packed around a pattern to form the mold. The pattern is removed, and on top of the cavity is placed a similar sand mold containing a passage (called a gate) through which the metal flows into the mold. The mold is designed so that solidification of the casting begins far from the gate and advances toward it, so that molten metal in the gate can flow in to compensate for the shrinkage that accompanies solidification. Sometimes additional spaces, called risers, are added to the casting to provide reservoirs to feed this shrinkage. After solidification is complete, the sand is removed from the casting, and the gate is cut off. If cavities are intended to be left in the casting—for example, to form hollow part— sand shapes called cores are made and suspended in the casting cavity before the metal is poured.

Patterns are also formed for sand-casting out of polymers that are evaporated by the molten metal. Such patterns may be injection molded and can possess a very complex shape. The process is called full-mold or evaporative pattern casting.

A variant of sand-casting is the shell-molding process, in which a mixture of sand and a thermosetting resin binder is placed on a heated metal pattern. The resin sets, binding the sand particles together and forming half of a strong mold. Two halves and any desired cores are then assembled to form the mold, and this mold is backed up with moist sand for casting. Greater dimensional accuracy and a smoother surface are obtained in this process than in greensand-casting.

Metal Molds

Other molds are made from metal. Here a die of the desired shape is machined from cast iron or steel. If the metal flows into the mold by gravity, the process is called permanent mold casting. If the molten metal is forced in under pressure, the process is called die casting. Die-casting dies are water-cooled; consequently, they can produce parts with thinner walls at a higher rate than permanent mold machines. The rapid cooling creates a stronger part than sand-casting, but ductility may be poorer owing to entrapped gas and porosity.

Since the initial cost of a die is substantial, metal molds are cost-effective only when many identical parts are to be made. Indeed, a die may be made to produce several parts at once.

Investment Casting

In investment casting a mold is made by drying refractory slurry on a pattern made of wax or plastic. A series of layers is applied and dried to make a ceramic shell, and the pattern is then melted or burned out to provide the mold. This process allows the mass production of parts with more complex shapes and finer surface detail than can be attained by other processes. It can be used with almost any metal and is customarily employed for casting relatively small parts. The wax pattern can be made by injection molding.

Centrifugal Casting

Centrifugal casting forces the metal into a mold by spinning it. It is used for the casting of small precious-metal objects, so that essentially all of the metal goes into the casting instead of the gates and risers. It is also used to produce long, hollow objects without resorting to cores—for example, to cast pipe. Here the long, cylindrical mold is horizontal and is spun about the axis of the cylinder as metal is poured into the mold.

Continuous Casting

Actually not a means of casting parts, continuous casting is practiced in the primary production of metals to form strands for further processing. The metal is poured into a short, reciprocating, water-cooled mold and solidifies even as it is withdrawn from the other side of the mold. The process is widely used in the steel industry because it eliminates the cost of reheating ingots and rolling them to the proportions of the billets, blooms, and slabs made by continuous casting.

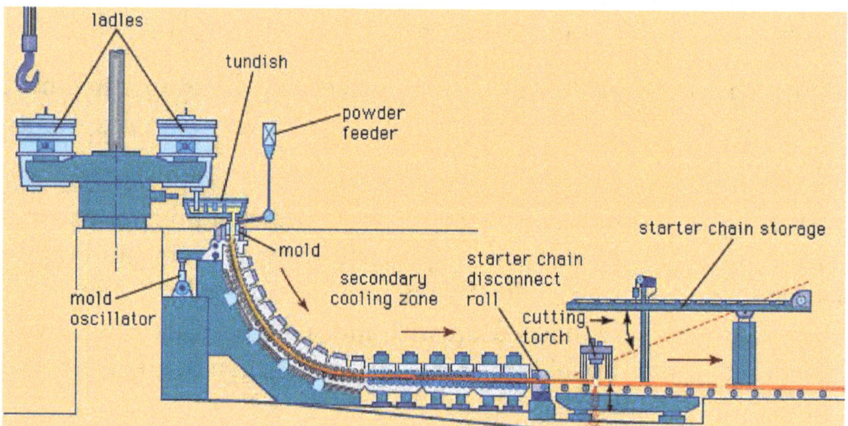

A curved-mold continuous slab caster.

Metallurgy

The mechanical properties of castings can be degraded by inhomogeneities in the solidifying metal. These include segregation, porosity, and large grain size.

Grain Size

A fine-grained casting can be produced by rapidly cooling the liquid metal to well below its equilibrium freezing temperature—*i.e.,* by pouring into a mold that cools the metal rapidly. For this reason, die castings have a finer grain size than the same alloy cast in a sand mold.

In cast iron, remarkable changes in microstructure result from various alloying additions and casting temperatures. For example, normal cast iron solidified in a sand mold forms what is known as gray iron, an iron matrix containing about 20 percent by volume graphite flakes. This type of iron has limited ductility. However, when a small amount of magnesium is added to the melt before casting, the result is a "spheroidal graphite" iron, in which graphite appears as spherical nodules and ductility is greatly increased. If the molten iron is chill cast (*i.e.*, rapidly cooled), it will form a "white" iron containing about 60 percent cementite, or iron carbide. This material is hard and wear-resistant, but it has no ductility at all. These cast irons are usually given a heat treatment to improve their mechanical properties.

Segregation

Different parts of a casting may have different compositions, stemming from the fact that the solid freezing out of a liquid has a different composition from the liquid with which it is in contact. (For example, when salt water is cooled until ice forms, the ice is essentially pure water while the salt concentration of the water rises.) Minor segregation is unimportant, but large differences can lead to local spots that are exceptionally weak or strong, and both of these can lead to early failure in a part under stress.

Porosity

A major problem in castings, porosity is principally caused by the shrinkage that accompanies solidification. Molds are designed to feed metal to the casting in order to keep it full as solidification proceeds, but, if this feeding is incomplete, the shrinkage will show up as internal pores or cracks. If these cracks are large, the casting will be useless. If they are small, they will have relatively little effect on the properties.

Another cause of porosity is the presence of gas-forming impurities in the liquid metal that exceed the solubility of the gas in the solid. In such cases, solidification is accompanied by the formation of bubbles as the gas is rejected. To eliminate this problem, gas-forming elements must be removed from the liquid before casting. Bubbling an inert gas such as argon through the liquid before casting is one means of doing this; vacuum degassing is another.

Metalworking

Processes

Metals are important largely because they can be easily deformed into useful shapes. Literally hundreds of metalworking processes have been developed for specific applications, but these can be divided into five broad groups: rolling, extrusion, drawing, forging, and sheet-metal forming. The first four processes subject a metal to large amounts of strain. However, if deformation occurs at a sufficiently high temperature, the metal will recrystallize—that is, its deformed grains will be consumed by the growth of a set of new, strain-free grains. For this reason, a metal is usually rolled, extruded, drawn, of forged above its recrystallization temperature. This is called hot working, and under these conditions there is virtually no limit to the compressive plastic strain to which the metal can be subjected.

Other processes are performed below the recrystallization temperature. These are called cold working. Cold working hardens metal and makes the part stronger. However, there is a definite limit to the strain that can be put into a cold part before it cracks.

Rolling

Rolling is the most common metalworking process. More than 90 percent of the aluminum, steel, and copper produced are rolled at least once in the course of production—usually to take the metal from a cast ingot down to a sheet or bar. The most common rolled product is sheet. With high-speed computer control, it is common for several stands of rolls to be combined in series, with thick sheet entering the first stand and thin sheet being coiled from the last stand at linear speeds of more than 100 kilometres (60 miles) per hour. Similar multistand mills are used to form coils of wire rod from bars. Other rolling mills can press large bars from several sides to form I-beams or railroad rails. Rolling can be done either hot or cold. If the rolling is finished cold, the surface will be smoother and the product stronger.

Extrusion

Extrusion converts a billet of metal into a length of uniform cross section by forcing the billet to flow through the orifice of a die. In forward extrusion the ram and the die are on opposite sides of the workpiece. Products may have either a simple or a complex cross section; examples of complex extrusions can be found in aluminum window frames.

Tubes or other hollow parts can also be extruded. The initial piece is a thick-walled tube, and the extruded part is shaped between a die on the outside of the tube and a mandrel held on the inside.

In impact extrusion (also called back-extrusion), the workpiece is placed in the bottom of a hole (the die), and a loosely fitting ram is pushed against it. The ram forces the metal to flow back around it, with the gap between the ram and the die determining the wall thickness. When toothpaste tubes were made of a lead alloy, they were formed by this process.

Drawing

Drawing consists of pulling metal through a die. One type is wire drawing. The diameter reduction that can be achieved in such a die is limited, but several dies in series can be used to obtain the desired reduction. Deep drawing starts with a disk of metal and ends up with a cup by pushing the metal through a hole (die). Several drawing operations in sequence may be used for one part. Deep drawing is employed in making aluminum beverage cans and brass rifle cartridges from sheet.

Sheet Metal Forming

In stretch forming, the sheet is formed over a block while the workpiece is under tension. The metal is stretched just beyond its yield point (2 to 4 percent strain) in order to retain the new shape.

Bending can be done by pressing between two dies. (Often a part can be made equally well by either stretch forming or bending; the choice then is made on the basis of cost.) Shearing is a cutting operation similar to that used for cloth. In these methods the thickness of the sheet changes little in processing.

Each of these processes may be used alone, but often all three are used on one part. For example, to make the roof of an automobile from a flat sheet, the edges are gripped and the piece pulled in tension over a lower die. Next a mating die is pressed over the top, finishing the forming operation, and finally the edges are sheared off to give the final dimensions.

Forging

Forging is the shaping of a piece of metal by pushing with open or closed dies. It is usually done hot in order to reduce the required force and increase the metal's plasticity.

Open-die forging is usually done by hammering a part between two flat faces. It is used to make parts that are too big to be formed in a closed die or in cases where only a few parts are to be made and the cost of a die is therefore unjustified. The earliest forging machines lifted a large hammer that was then dropped on the work, but now air- or steam-driven hammers are used, since these allow greater control over the force and the rate of forming. The part is shaped by moving or turning it between blows. A forged ring can be formed by placing a mandrel through the ring and deforming the metal between the hammer and the mandrel. Rings also can be forged by rolling with one roll inside the ring and the other outside.

Closed-die forging is the shaping of hot metal within the walls of two dies that come together to enclose the work piece on all sides. The process starts with a rod or bar cut to the length needed to fill the die. Since large, complex shapes and large strains are involved, several dies may be used to go from the initial bar to the final shape. With closed dies, parts can be made to close tolerances so that little finish machining is required.

Two closed-die forging operations given special names are upsetting and coining. Coining takes its name from the final stage of forming metal coins, where the desired imprint is formed on a smooth metal disk that is pressed in a closed die. Coining involves small strains and is done cold to enhance surface definition and smoothness. Upsetting involves a flow of the metal back upon itself. An example of this process is the pushing of a short length of a rod through a hole, clamping the rod, and then hitting the exposed length with a die to form the head of a nail or bolt.

Metallurgy

An important benefit of hot working is that it provides control over and improvement of mechanical properties. Hot-rolling or hot-forging eliminate much of the porosity, directionality, and segregation that may be present in cast shapes. The resulting "wrought" product therefore has better ductility and toughness than the unworked casting. During the forging of a bar, the grains of the metal become greatly elongated in the direction of flow. As a result, the toughness of the metal is substantially improved in this direction and somewhat weakened in directions transverse to the flow. Part of the design of a good forging is to assure that the flow lines in the finished part are oriented so as to lie in the direction of maximum stress when the part is placed in service.

The ability of a metal to resist thinning and fracture during cold-working operations plays an important role in alloy selection and process design. In operations that involve stretching, the best alloys are those which grow stronger with strain (strain harden)–for example, the copper-zinc alloy, brass, used for cartridges and the aluminum-magnesium alloys in beverage cans, which exhibit greater strain hardening than do pure copper or aluminum, respectively.

Another useful property that can be controlled by processing and composition is the plastic anisotropy ratio. When a segment of sheet is strained (*i.e.*, elongated) in one direction, the thickness and width of the segment must shrink, since the volume remains constant. In an isotropic sheet the thickness and width show equal strain, but, if the grains of the sheet are oriented properly, the thickness will shrink only about half as much as the width. Since it is thinning that leads to early fracture, this plastic anisotropy imparts better deep-drawing properties to sheet material with the optimum grain orientation.

Fracture of the workpiece during forming can result from flaws in the metal; these often consist of nonmetallic inclusions such as oxides or sulfides that are trapped in the metal during refining. Such inclusions can be avoided by proper manufacturing procedures. Laps are another type of flaw in which part of a metal piece is inadvertently folded over on itself but the two sides of the fold are not completely welded together. If a force tending to open this fold is applied during the forming operation, the metal will fail at the lap.

The ability of different metals to undergo strain varies appreciably. The shape change that can be made in one forming operation is often limited by the tensile ductility of the metal. Metals with the face-centred cubic crystal structure, such as copper and aluminum, are inherently more ductile in such operations than metals with the body-centred cubic structure. To avoid early fracture in the latter type of metals, processes are used that apply primarily compressive stresses rather than tensile stresses.

Powder Metallurgy

Powder metallurgy (P/M) consists of making solid parts out of metal powders. The powder is mixed with a lubricant, pressed into a die to form the desired shape, and then sintered, or heated to a temperature below the melting point of the alloy where solid-state bonding of the particles takes place. In the absence of any external force, sintering typically leaves the sample containing about 5 percent pores by volume, but, when pressure is applied during sintering (a process called hot pressing), virtually zero porosity remains. In some parts made by mixing two different elements, one component melts at the sintering temperature and this liquid phase aids sintering of the solid particles.

Applications

The following roughly chronological account indicates the types of products that can be made by P/M.

The earliest commercial use of P/M was in the production of such high-melting-point metals as platinum, tungsten, and tantalum. Pure powders of these metals could be made by the low-temperature reduction of powders, usually oxides, and, since these metals melt at extremely high temperatures, it was easier to form solid parts by pressing and sintering the powders than by melting and casting. For example, P/M played an important role in the development of tungsten filaments for electric light bulbs.

Another early P/M product was porous-metal bearings and filters. In such parts sintering is conducted at a relatively low temperature so that the pores between the particles remain open and connected. Disks sintered in this way can serve as filters for liquids, or the sintered part can be impregnated with oil to make a self-lubricating bearing. In the latter case, the oil is held in the pores by surface tension. When the bearing heats up in use, some oil flows out and lubricates the surface, and, when the part cools, surface tension pulls the oil back into the fine channels.

Cemented carbides form another class of sintered product. Pure tungsten carbide (WC) is an extremely hard compound, but it is too brittle to be used in tools. However, useful tools can be made by mixing WC powder with cobalt powder and sintering at a temperature above the melting point of cobalt. Liquid cobalt then reacts with the surface of the WC, and when the part is cooled the cobalt freezes, holding the WC tightly together to form a composite structure with enough toughness to be used for tools and dies.

The greatest volume of P/M parts is now produced from iron powder, a process that was first developed during World War II. Small, complex parts, such as gears, require much work if machined from steel bars, and a significant volume of material is wasted as chips from the machining. However, if the part is made by P/M processes, little or no machining is necessary, there is less wasted material, and the cost is much lower. Many small parts for automobiles and appliances are produced in this manner. The second greatest volume of P/M parts is made from aluminum powder. These parts are light, corrosion-resistant, and (if alloy is used) can be heat treated to appreciably increase the strength. Small parts for automobiles and appliances are the most common applications.

A recent process uses P/M methods to improve the homogeneity and toughness of high-alloy tool steels. Cast ingots of these alloys contain a coarse network of brittle phases that are very difficult to break up by hot working, but if, instead of being cast into ingots, the liquid is atomized (solidified as small droplets), the rapidly solidified particles will be homogeneous. This powder can then be hot pressed into consolidated bars with better mechanical properties than those produced by ingot casting. Consolidation is often achieved by hot isostatic pressing, wrapping the pressed powder in an envelope of steel or glass, and heating it in a hot inert gas at high pressure. The consolidated metal is then worked into finished parts.

Processes

The most common method of producing metal powders is atomization of a liquid. Here a stream of molten metal is broken up into small droplets with a jet of water, air, or inert gas such as nitrogen or argon. Atomization in water yields irregularly shaped particles that can be pressed to a higher initial, or "green," strength and density than can the spherical particles formed by atomizing with an inert gas.

In other atomization processes, centrifugal force is used. The metal can be poured onto a spinning disk that breaks up the stream, or a spinning rod can be melted by an electric arc so that it throws off particles as it spins.

The liquid metal being atomized may be an alloy or a pure metal that will subsequently be blended with other elements to form an alloy. After atomization, the powder must be separated into size ranges by passing it through a series of sieves. Powders of different sizes (and of different metals) are then blended for pressing into parts.

Powders are often produced by chemically reducing a powdered oxide—for example, iron oxide reduced with carbon or hydrogen. The resulting metal aggregate is then milled and sieved to obtain the desired powder. Powder also can be made by electrodepositing the metal at a high current density, followed by milling to break up the deposit.

The above processes often produce powders that are roughly 50 to 200 micrometres in diameter. Powders less than one-tenth this size can be found in the finest fraction of the powder produced by atomization. Such fine powders can be mixed with a wax, injection molded to form several parts at once, and then sintered. The resulting parts require very little machining to yield a finished product.

Heat Treating

The properties of metals can be substantially changed by various heat-treatment processes. Depending on the alloy and its condition, heat treating can harden or soften a metal.

Hardening Treatments

Hardening heat treatments invariably involve heating to a sufficiently high temperature to dissolve solute-rich precipitates. The metal is then rapidly cooled to avoid re-precipitation; often this is done by quenching in water or oil. The concentration of solute dissolved in the metal is now much greater than the equilibrium concentration. This produces what is known as solid-solution hardening, but the alloy can usually be hardened appreciably more by aging to allow a very fine precipitate to form. Aging is done at an elevated temperature that is still well below the temperature at which the precipitate will dissolve. If the alloy is heated still further, the precipitate will coarsen; that is, the finest particles will dissolve so that the average particle size will increase. This will reduce the hardness somewhat but increase the ductility. Precipitation hardening is used to produce most high-strength alloys. In products made of soft, ductile metals such as aluminum or copper, the age-hardened alloy is put into service with the finest precipitate (that is, the highest strength) possible.

When heated to a high temperature, a few metals principally iron and titanium alloys, transform to a different crystal structure. Often the high-temperature phase has a higher solubility for the solute, thus aiding the dissolution of the precipitate. If the alloy is slowly cooled, the reverse phase transformation will occur at a high temperature, forming a coarse precipitate and yielding a soft structure. This is the principle behind annealing procedures. However, if the alloy is quenched from the high temperature, the reverse transformation occurs at a much lower temperature, so that a very fine precipitate forms. This is the basis for hardening iron-carbon (steel) alloys. The hardness of the low-temperature-transformation phase (known as martensite) increases with carbon content, and this can result in some very strong alloys. Other alloying elements such as nickel, chromium, and manganese are added to steel primarily to slow transformation from the high-temperature phase so that thicker pieces, which cool more slowly on quenching, will harden to martensite on cooling. Steel with fresh martensite is still not tough enough for use without first being heated to an elevated temperature. This tempering process reduces residual stresses produced by the phase transformation, reduces hardness by coarsening the carbide precipitate, and increases toughness. Where high strength is the main concern, the tempering temperature is kept low. When toughness is the primary goal and strength secondary, a relatively high tempering temperature is used.

Softening Treatments

In many situations the purpose of heat treating is to soften the alloy and thereby increase its ductility. This may be necessary if a number of cold-forming operations are required to form a part but the metal is so hardened after the first operation that further cold working will cause it to crack. If the metal is recrystallized by annealing at an elevated temperature, it will become soft enough to allow further forming operations. Another case arises when it is necessary to machine high-carbon tools in order to form a die. If the alloy is quenched from a high temperature to form martensite, it will be hard, brittle, and virtually impossible to machine. But if it is slowly cooled, the carbides will be much coarser, and the steel will be machinable.

Processes

Most furnaces designed for heat treatment use natural gas or electricity to raise the temperature. The atmosphere around the work may be air for low-temperature anneals, but at elevated temperatures some atmosphere other than air must be used in order to avoid oxidation. One common atmosphere is obtained by burning natural gas with less than the stoichiometric amount of air. With more reactive metals, annealing can be done in a vacuum furnace.

In some heat treatments only the surface need be heated. With electromagnetic induction or by use of a laser, this can be done so quickly that no special atmosphere is needed to avoid oxidation. Surface heat treating also avoids the distortion that can accompany heating and quenching the entire part. For example, the rear axle of most automobiles is a steel bar roughly 1 metre long and 3 centimetres in diameter (about 3 feet long and 1.25 inches in diameter). The surface can be hardened by passing the bar through an induction coil that quickly heats the surface immediately beneath the coil to red heat, transforming it to austenite. The inside remains cold, however, and, after the coil passes, this cold interior quickly draws heat from the surface, transforming it to martensite. The part is then tempered and put into service. Since most of the cross section of the part is not heated or transformed during this operation, there is no distortion and therefore no need to straighten a hardened part.

Surface Treating

Because it is the surface of a metal that people see and that reacts with the environment around it, special effort is sometimes made to add lustre, colour, or texture to a surface. In addition, special corrosion-resistant layers are placed on the surface for some applications, and in yet others the surface is hardened to add strength and reduce wear.

Angle grinder: Worker using an angle grinder to smooth a piece of metal.

Corrosion Resistance

Barrier Protection

When a metal corrodes in water, the atoms lose electrons and become ions that move into the water. This is called an anodic reaction, and for the corrosion process to proceed there must be a corresponding cathodic reaction that adsorbs the electrons. The process can be stopped by isolating the metal from the water with an impermeable barrier. One of the older applications of this idea is the tin can. Unlike steel, tin is not affected by the acids in food, so that a layer of tin placed on steel sheet protects the steel in the can from corrosion.

The exterior surfaces of many large household appliances consist of steel covered with a layer of coloured glass called enamel. Enamel is inert and adheres tightly to the steel, thus protecting it from corrosion as well as providing an attractive appearance. Decorative chromium plating is another example of a protective-barrier coating on steel. Since chromium does not adhere well to steel, the steel is first electroplated with layers of copper and nickel before being plated with a thin layer of chromium.

The protective layers described above are metallic, but the most common protective barriers are organic. Paints, polymers, and thin lacquer films are used for various applications near room temperature.

The oxide layer that forms on metals when they are exposed to air also constitutes a protective barrier. Stainless steel and aluminum form the most stable and protective of such films. The thickness of the oxide film on aluminum is often increased by making the part function as the anode in an electrolytic cell. This process, called anodizing, enhances the corrosion resistance somewhat and makes colouring the surface easier. The films that form on copper and steel as a result of corrosion (commonly known as tarnish and rust) are somewhat thicker and show a characteristic colour that is often incorporated into the design of the part.

Galvanic Protection

Protective films on steel are susceptible to being broken by flying stones or other sharp objects. This is especially true on the parts of an automobile that are close to the road. The barriers described above have limited ability for self-healing after they are broken, but protection of the underlying small exposed area of steel can be maintained if a metal that has a greater tendency than steel to give up electrons in water is attached to the surface. In this way, when the protective barrier is broken, the more reactive metal is corroded preferentially, and the steel is given "galvanic" protection.

A layer of zinc can be placed on a steel surface either by hot-dipping the steel in molten zinc or by electroplating zinc onto the surface. Galvanized steel is much more resistant to corrosion than ungalvanized steel. For this reason, it is often used in the lowest panels of automobiles—*i.e.*, the parts exposed to corrosive salt spray from snowy roads.

Electrodeposits of cadmium are also used for galvanic protection of steel. Hot-dip aluminum-coated steel is used in the exhaust systems of automobiles. At low temperatures its action is primarily galvanic, while at high temperatures it oxidizes to form a barrier layer.

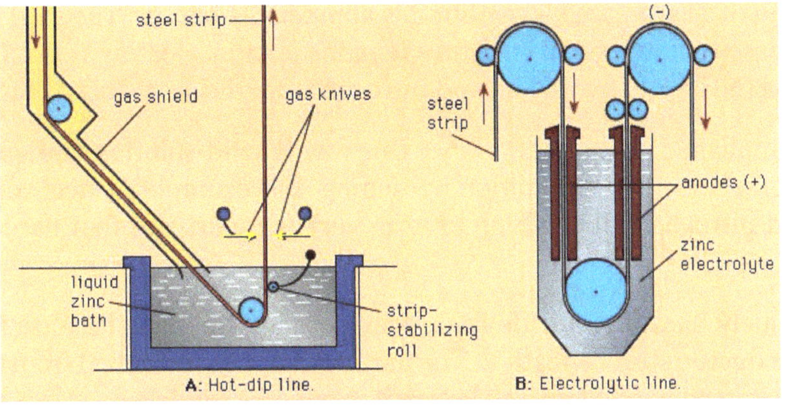

Principles of (A) hot-dip and (B) electrolytic galvanizing.

Other Coating Techniques

Among other methods for applying metal layers to metal is thermal spray coating, a generic term for processes in which a metal wire is melted by a plasma arc or a flame, atomized, and sprayed onto a surface in an inert gas. Another method is vacuum coating, which produces thin, bright, attractive layers on a part by evaporating and depositing a coating metal in a high vacuum.

Enhanced oxidation protection is imparted to high-temperature turbine parts made of superalloys by annealing the parts in a container containing volatile aluminum chloride. This is done at high temperatures, so that aluminum diffuses into the alloy to form an aluminum-rich surface layer. In high-temperature service, such a layer oxidizes to form a protective aluminum oxide coating.

Hardening

Carburizing

The strength of hardened steel increases rapidly as the percentage of carbon is increased, but at the same time the steel's toughness decreases. Often the most useful part is one in which the surface is higher in carbon and thus hard, while the interior is lower in carbon and thus tough. Such a combination of properties can be obtained by carburizing, or annealing the parts in a gas rich in carbon. (The carburizing potential of the gas rises with the ratio of carbon monoxide to carbon dioxide.) The carburizing temperature is high enough to transform the surface of the steel to the high-temperature austenite phase, which has much higher carbon solubility than the low-temperature ferrite phase. At these temperatures, carbon deposited on the surface diffuses through the steel and into the solid. The thickness of the diffusion layer increases with time, although at a decreasing rate; depths of 1 to 2 millimetres (0.04 to 0.08 inch) in 4 to 16 hours are typical.

Following diffusion, the part is quenched in oil. The high-carbon surface transforms into a hard, brittle martensitic structure, while the lower-carbon interior transforms into a tougher, softer structure. The part is then tempered to raise the toughness of the surface layer. Small machine parts, such as gears, are often carburized to increase their strength and resistance to wear.

Nitriding

Nitriding provides an alternative means of hardening a steel surface. The surface layer is only one-tenth the depth of a carburized layer, but it is appreciably harder. The steel part is heated to a lower temperature, so that its crystal structure remains ferritic. Heating is conducted in an atmosphere of ammonia (NH_3) and hydrogen, and nitrogen from the ammonia diffuses into the steel.

Hardening is accomplished in one of two ways. One way is solid-solution hardening, which occurs in all steels. The other way is precipitation hardening. For example, if steel contains aluminum, the aluminum and nitrogen will combine to form very fine particles that harden the steel quite effectively.

Though it is extremely hard, the nitride layer does not tend to crack, because it is very thin and adheres well to the ductile steel beneath it. The part need not be quenched from the nitriding temperature, nor does it need to be tempered before it is put into service.

Other Methods

Surfaces can also be hardened by heat treating with induction or laser heating. In other applications, surface hardening is accomplished by depositing a "hard-facing alloy" on the base part. One example is "hard chromium plating," in which a thick layer of chromium is deposited on a part. Automotive valve stems and piston rings and the bores of diesel engine cylinders are common applications.

Metallography and Metal Testing

Metallography

The properties of an alloy of a given composition can change markedly with the microscopic arrangement of its crystalline grains—*i.e.*, its microstructure. To evaluate and control the microstructure of a sample, various types of microscope are used, and the field is called metallography.

Optical Microscopy

The simplest, and oldest, type of metallography (though hardly a century old) involves polishing the surface to a mirror like finish and examining light reflected from it at magnifications of 50 to 1500×. If the surface is lightly etched in an appropriate solution (often an acid), the grain boundaries, matrix, and constituent phases will be attacked at different rates and will be discernible. This makes it possible to establish which phases are present as well as their shape, size, and distribution. Similarly, grain size and shape can be observed. With this information it is possible to infer the history of the sample and predict its behaviour. Metallography is of particular value in the analysis of samples that have failed or performed in an unexpected manner.

Electron Microscopy

Great progress has been made in using finely focused beams of energetic electrons to examine metals. Electron microscopes are basically of two types, transmission and scanning. Transmission electron microscopes require the preparation of films so thin that they are transparent to a beam of electrons with energies of roughly 200 kiloelectron volts. This means the film must have a thickness of only one, or a few, hundred nanometres (10^{-9} metre). Films of lighter elements, such as aluminum, can be thicker, while films of heavier elements, such as gold, must be thinner. Contrast between neighbouring regions is best developed by differences in their diffraction of the electron beam, although differences in density can also be used. Spatial resolution is excellent, going down to atomic resolution in special microscopes, and orientation relations between adjoining regions can be easily discerned. On the other hand, only very small samples can be examined in any given film. This means that the technique is not good for measuring defects larger than the film thickness or those whose number per unit volume is low.

A scanning electron microscope (SEM) uses a narrow beam of electrons (often of about 40 kiloelectron volts) that scans the surface of a sample and forms a corresponding image from the backscattered electrons or secondary electrons. No special surface preparation is necessary, and, since the depth of focus in an SEM is much greater than in an optical microscope, quite irregular surfaces, such as fractures, can be studied successfully. (The strikingly detailed pictures of insects seen in publications are also taken with an SEM.) Useful magnifications range from 100 to 20,000×.

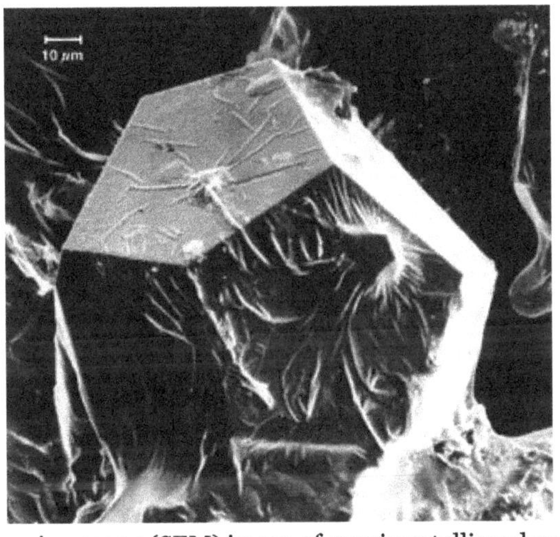

A scanning electron microscope (SEM) image of quasicrystalline aluminum-copper-iron, revealing the pentagonal dodecahedral shape of the grain.

The electron beam used in an SEM causes each atom near the surface to emit an X ray that is characteristic of that element. By constructing an image based on the distribution of the intensity of the characteristic X ray of a given element, it is possible to show that element's distribution among the phases in the surface. If the electron beam is not swept but held in one spot, a chemical analysis can be made of the various elements in the region under the electron beam by measuring the intensity of the X rays emitted by each element.

Polymer Science and Engineering

Polymer Science

Polymer science is an interdisciplinary area comprised of chemical, physical, engineering, processing and theoretical aspects. It also has enormous impact on contemporary materials science. Its goal is to provide the basis for the creation and characterization of polymeric materials and an understanding for structure/property relationships. Polymer science is of increasing importance for everyone's daily life. Many modern functional materials, gears, and devices have polymers as integral parts.

Polymer Engineering

Polymer engineering is generally an engineering field that designs, analyses, and modifies polymer materials. Polymer engineering covers aspects of the petrochemical industry, polymerization, structure and characterization of polymers, properties of polymers, compounding and processing of polymers and description of major polymers, structure property relations and applications.

Polymer Materials used in Engineering

Plastic is a polymer compound which is polymerized by polyaddition polymerization and polycondensation. It is free to change the composition and shape. It is made up of synthetic resins and fillers, plasticizers, stabilizers, lubricants, colorants and other additives. The main component of plastic is resin. Resin means that the polymer compound has not been added with various additives. The term resin was originally named for the secretion of oil from plants and animals, such as rosin and shellac. Resin accounts for approximately 40% - 100% of the total weight of the plastic. The basic properties of plastics are mainly determined by the nature of the resin, but additives also play an important role. Some plastics are basically made of synthetic resins, with or without additives such as plexiglass, polystyrene, etc.

Fiber

Fiber refers to a continuous or discontinuous filament of one substance. Animals and plant fibers play an important role in maintaining tissue. Fibers are widely used and can be woven into good threads, thread ends and hemp ropes. They can also be woven into fibrous layers when making paper or feel. They are also commonly used to make other materials together with other materials to form composites. Therefore, whether it is natural or synthetic fiber filamentous material. In modern life, the application of fiber is ubiquitous, and there are many high-tech products.

Rubber

Rubber refers to highly elastic polymer materials and reversible shapes. It is elastic at room temperature and can be deformed with a small external force. After removing the external force, it can return to the original state. Rubber is a completely amorphous polymer with a low glass transition temperature and a large molecular weight, often greater than several hundred thousand. Highly elastic polymer compounds can be classified into natural rubber and synthetic rubber. Natural rubber processing extracts gum rubber and grass rubber from plants; synthetic rubber is polymerized by various monomers. Rubber can be used as elastic, insulating, water-impermeable air-resistant materials.

Applications

Polyethylene

Commonly used polyethylenes can be classified into low density polyethylene (LDPE), high density polyethylene (HDPE), and linear low density polyethylene (LLDPE). Among them, HDPE has better thermal, electrical and mechanical properties, while LDPE and LLDPE have better flexibility, impact properties and film forming properties. LDPE and LLDPE are mainly used for plastic bags, plastic wraps, bottles, pipes and containers; HDPE is widely used in various fields such as film, pipelines and daily necessities because its resistance to many different solvents.

B-2 Spirit stealth bomber of the U.S. Air Force.

Polypropylene

Polypropylene is widely used in various applications due to its good chemical resistance and weldability. It has lowest density among commodity plastics. It is commonly used in packaging applications, consumer goods, automatic applications and medical applications. Polypropylene sheets are widely used in industrial sector to produce acid and chemical tanks, sheets, pipes, Returnable Transport Packaging (RTP), etc. because of its properties like high tensile strength, resistance to high temperatures and corrosion resistance.

Composites

A time-trial carbon fibre composite bicycle with aerodynamic wheels and aero bars.

Typical uses of composites are monocoque structures for aerospace and automobiles, as well as more mundane products like fishing rods and bicycles. The stealth bomber was the first all-composite aircraft, but many passenger aircraft like the Airbus and the Boeing 787 use an increasing proportion of composites in their fuselages, such as hydrophobic melamine foam. The quite different physical properties of composites gives designers a much greater freedom in shaping parts, which is why composite products often look different to conventional products. On the other hand, some products such as drive shafts, helicopter rotor blades, and propellers look identical to metal precursors owing to the basic functional needs of such components.

Biomedical Applications

Biodegradable polymers are widely used materials for many biomedical and pharmaceutical applications. They are considered vey promising for controlled drug delivery devices. Biodegradable polymers also offer great potential for wound management, orthopaedic devices, dental applications and tissue engineering. Not like non-biodegradable polymers, they won't require a second step of a removal from body. Biodegradable polymers will break down and are absorbed by the body after they served their purpose. Since 1960, polymers prepared from glycolic acid and lactic acid have found a multitude of uses in the medical industry. Polylactates (PLAs) are popular for drug delivery system due to their fast and adjustable degradation rate.

Membrane Technologies

Membrane techniques are successfully used in the separation in the liquid and gas systems for years, and the polymeric membranes are used most commonly because they have lower cost to produce and are easy to modify their surface, which make them suitable in different separation processes. Polymers help in many fields including the application for separation of biological active compounds, proton exchange membranes for fuel cells and membrane contractors for carbon dioxide capture process.

Nanotechnology

Nanotechnology is the manipulation and manufacture of materials and devices on the scale of atoms or small groups of atoms. The "nanoscale" is typically measured in nanometres, or billionths of a metre and materials built at this scale often exhibit distinctive physical and chemical properties due to quantum mechanical effects. Although usable devices this small may be decades away, techniques for working at the nanoscale have become essential to electronic engineering, and nanoengineered materials have begun to appear in consumer products. For example, billions of microscopic "nanowhiskers," each about 10 nanometres in length, have been molecularly hooked onto natural and synthetic fibres to impart stain resistance to clothing and other fabrics; zinc oxide nanocrystals have been used to create invisible sunscreens that block ultraviolet light; and silver nanocrystals have been embedded in bandages to kill bacteria and prevent infection.

Possibilities for the future are numerous. Nanotechnology may make it possible to manufacture lighter, stronger, and programmable materials that require less energy to produce than conventional materials, that produce less waste than with conventional manufacturing, and that promise greater fuel efficiency in land transportation, ships, aircraft, and space vehicles. Nanocoatings for both opaque and translucent surfaces may render them resistant to corrosion, scratches, and radiation.

Nanoscale electronic, magnetic, and mechanical devices and systems with unprecedented levels of information processing may be fabricated, as may chemical, photochemical, and biological sensors for protection, health care, manufacturing, and the environment; new photoelectric materials that will enable the manufacture of cost-efficient solar-energy panels; and molecular-semiconductor hybrid devices that may become engines for the next revolution in the information age. The potential for improvements in health, safety, quality of life, and conservation of the environment are vast.

At the same time, significant challenges must be overcome for the benefits of nanotechnology to be realized. Scientists must learn how to manipulate and characterize individual atoms and small groups of atoms reliably. New and improved tools are needed to control the properties and structure of materials at the nanoscale; significant improvements in computer simulations of atomic and molecular structures are essential to the understanding of this realm. Next, new tools and approaches are needed for assembling atoms and molecules into nanoscale systems and for the further assembly of small systems into more-complex objects. Furthermore, nanotechnology products must provide not only improved performance but also lower cost. Finally, without integration of nanoscale objects with systems at the micro- and macroscale (that is, from millionths of a metre up to the millimetre scale), it will be very difficult to exploit many of the unique properties found at the nanoscale.

Nanotechnology is highly interdisciplinary, involving physics, chemistry, biology, materials science, and the full range of the engineering disciplines. The word *nanotechnology* is widely used as shorthand to refer to both the science and the technology of this emerging field. Narrowly defined, nanoscience concerns a basic understanding of physical, chemical, and biological properties on atomic and near-atomic scales. Nanotechnology, narrowly defined, employs controlled manipulation of these properties to create materials and functional systems with unique capabilities.

In contrast to recent engineering efforts, nature developed "nanotechnologies" over billions of years, employing enzymes and catalysts to organize with exquisite precision different kinds of atoms and molecules into complex microscopic structures that make life possible. These natural products are built with great efficiency and have impressive capabilities, such as the power to harvest solar energy, to convert minerals and water into living cells, to store and process massive amounts of data using large arrays of nerve cells, and to replicate perfectly billions of bits of information stored in molecules of deoxyribonucleic acid (DNA).

There are two principal reasons for qualitative differences in material behaviour at the nanoscale (traditionally defined as less than 100 nanometres). First, quantum mechanical effects come into play at very small dimensions and lead to new physics and chemistry. Second, a defining feature at the nanoscale is the very large surface-to-volume ratio of these structures. This means that no atom is very far from a surface or interface, and the behaviour of atoms at these higher-energy sites have a significant influence on the properties of the material. For example, the reactivity of a metal catalyst particle generally increases appreciably as its size is reduced—macroscopic gold is chemically inert, whereas at nanoscales gold becomes extremely reactive and catalytic and even melts at a lower temperature. Thus, at nanoscale dimensions material properties depend on and change with size, as well as composition and structure.

Using the processes of nanotechnology, basic industrial production may veer dramatically from the course followed by steel plants and chemical factories of the past. Raw materials will come from the atoms of abundant elements—carbon, hydrogen, and silicon—and these will be manipulated into precise configurations to create nanostructured materials that exhibit exactly the right properties for each particular application. For example, carbon atoms can be bonded together in a number of different geometries to create variously a fibre, a tube, a molecular coating, or a wire, all with the superior strength-to-weight ratio of another carbon material—diamond. Additionally, such material processing need not require smokestacks, power-hungry industrial machinery, or intensive human labour. Instead, it may be accomplished either by "growing" new structures through some combination of chemical catalysts and synthetic enzymes or by building them through new techniques based on patterning and self-assembly of nanoscale materials into useful predetermined designs. Nanotechnology ultimately may allow people to fabricate almost any type of material or product allowable under the laws of physics and chemistry. While such possibilities seem remote, even approaching nature's virtuosity in energy-efficient fabrication would be revolutionary.

Even more revolutionary would be the fabrication of nanoscale machines and devices for incorporation into micro- and macroscale systems. Once again, nature has led the way with the fabrication of both linear and rotary molecular motors. These biological machines carry out such tasks as muscle contraction (in organisms ranging from clams to humans) and shuttling little packets of material around within cells while being powered by the recyclable, energy-efficient fuel adenosine triphosphate. Scientists are only beginning to develop the tools to fabricate functioning systems at such small scales, with most advances based on electronic or magnetic information processing and storage systems. The energy-efficient, reconfigurable, and self-repairing aspects of biological systems are just becoming understood.

The potential impact of nanotechnology processes, machines, and products is expected to be far-reaching, affecting nearly every conceivable information technology, energy source, agricultural product, medical device, pharmaceutical, and material used in manufacturing. Meanwhile, the dimensions of electronic circuits on semiconductors continue to shrink, with minimum feature sizes now reaching the nanorealm, under 100 nanometres. Likewise, magnetic memory materials, which form the basis of hard disk drives, have achieved dramatically greater memory density as a result of nanoscale structuring to exploit new magnetic effects at nanodimensions. These latter two areas represent another major trend, the evolution of critical elements of microtechnology into the realm of nanotechnology to enhance performance. They are immense markets driven by the rapid advance of information technology.

Nanomaterials

Nanomaterials are cornerstones of nanoscience and nanotechnology. Nanostructure science and technology is a broad and interdisciplinary area of research and development activity that has been growing explosively worldwide in the past few years. It has the potential for revolutionizing the ways in which materials and products are created and the range and nature of functionalities that can be accessed. It is already having a significant commercial impact, which will assuredly increase in the future.

Nanoscale materials are defined as a set of substances where at least one dimension is less than approximately 100 nanometers. A nanometer is one millionth of a millimeter - approximately 100,000 times smaller than the diameter of a human hair. Nanomaterials are of interest because at this scale unique optical, magnetic, electrical, and other properties emerge. These emergent properties have the potential for great impacts in electronics, medicine, and other fields.

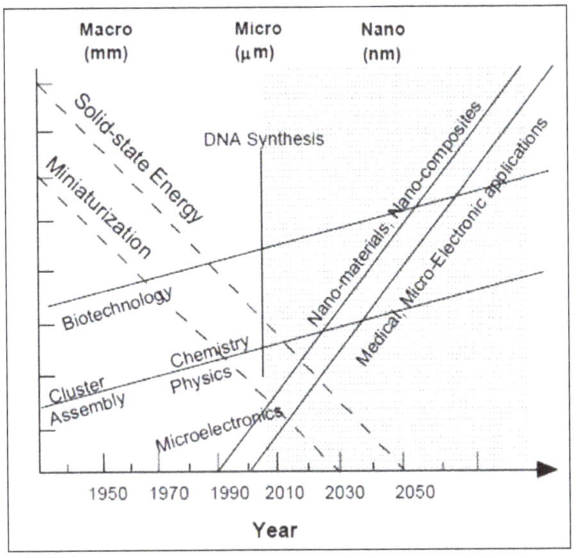

Evolution of science & technology and the future.

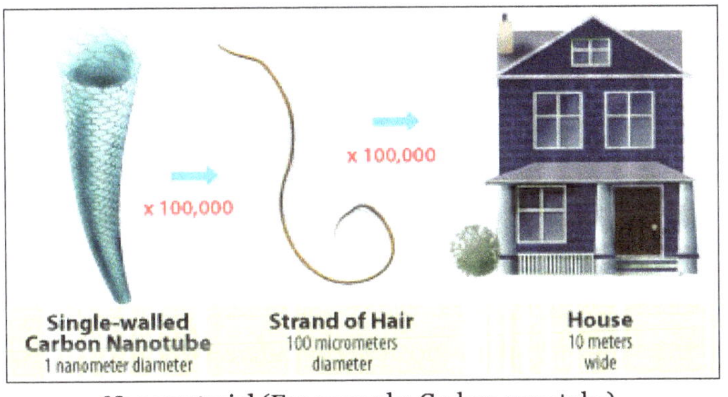

Nanomaterial (For example: Carbon nanotube).

Occurence of Nanomaterials

Some nanomaterials occur naturally, but of particular interest are engineered nanomaterials (EN), which are designed for, and already being used in many commercial products and processes. They can be found in such things as sunscreens, cosmetics, sporting goods, stain resistant clothing, tires, electronics, as well as many other everyday items, and are used in medicine for purposes of diagnosis, imaging and drug delivery.

Engineered nanomaterials are resources designed at the molecular (nanometre) level to take advantage of their small size and novel properties which are generally not seen in their conventional, bulk counterparts. The two main reasons why materials at the nano scale can have different properties are increased relative surface area and new quantum effects. Nanomaterials have a much greater surface area to volume ratio than their conventional forms, which can lead to greater chemical reactivity and affect their strength. Also at the nano scale, quantum effects can become much more important in determining the materials properties and characteristics, leading to novel optical, electrical and magnetic behaviours.

Nanomaterials are already in commercial use, with some having been available for several years or decades. The range of commercial products available today is very broad, including stainresistant and wrinkle-free textiles, cosmetics, sunscreens, electronics, paints and varnishes. Nanocoatings and nanocomposites are finding uses in diverse consumer products, such as windows, sports equipment, bicycles and automobiles. There are novel UV-blocking coatings on glass bottles which protect beverages from damage by sunlight, and longer-lasting tennis balls using butyl-rubber/nano-clay composites. Nanoscale titanium dioxide, for instance, is finding applications in cosmetics, sun-block creams and self-cleaning windows, and nanoscale silica is being used as filler in a range of products, including cosmetics and dental fillings.

Nanomaterial Applications using Carbon Nanotubes

Applications being developed for carbon nanotubes include adding antibodies to nanotubes to form bacteria sensors, making a composite with nanotubes that bend when electric voltage is applied bend the wings of morphing aircraft, adding boron or gold to nanotubes to trap oil spills, include smaller transistors, coating nanotubes with silicon to make anodes the can increase the capacity of Li-ion batteries by up to 10 times. Check out our Applications of Carbon Nanotubes page to see how carbon nanotubes are being used.

Nanomaterial Applications using Graphene

Applications being developed for graphene include using graphene sheets as electodes in ultra-capacitors which will have as much storage capacity as batteries but will be able to recharge in minutes, attaching strands of DNA to graphene to form sensors for rapid disease diagnostics, replacing indium in flat screen TVs and making high strenght composite materials.

Nanomaterial Applications using Nanocomposites

Applications being developed for nanocomposites include a nanotube-polymer nanocomposite to form a scaffold which speeds up replacement of broken bones, making a graphene-epoxy nanocomposite with very high strenght-to-weight ratios, a nanocomposite made from cellulous and nanotubes used to make a flexible battery.

Nanomaterial Applications using Nanofibers

Applications being developed for nanofibers include stimulating the production of cartilage in damaged joints, piezoelectric nanofibers that can be woven into clothing to produce electricty for cell phones or other devices, carbon nanofibers that can improve the preformance flame retandant in funiture.

Nanomaterial Applications using Nanoparticles

Applications being developed for nanoparticles include deliver chemotherapy drugs directly to cancer tumors, resetting the immune system to prevent autoimmune diseases, delivering drugs to damaged regions of arteries to fight cardiovascular disease, create photocatalysts that produce hydrogen from water, reduce the cost of producing fuel cells and solar cells, clean up oil spills, water pollution and air pollution.

Nanomaterial Applications using Nanowires

Applications being developed for carbon nanotubes include using zinc oxide nanowires in a flexible solar cell, silver chloride nanowires to decompose organic molecules in polluted water, using nanowires made from iron and nickel to make dense computer memory - called “race track memory.

Crystallography

Crystallography is the experimental science of determining the arrangement of atoms in solids. In older usage, it referred to the scientific study of crystals.

Before the development of X-ray diffraction crystallography, the study of crystals was based on their geometry. It involved measuring the angles of crystal faces relative to theoretical reference axes (crystallographic axes), using a goniometer, and establishing the symmetry of the crystal in question.

Crystallographic methods now depend on the analysis of the diffraction patterns that emerge from a sample that is targeted by a beam of some type. The beam is not always electromagnetic radiation, although X-rays are the most common choice. For some purposes, electrons or neutrons are used, which is possible due to the wave properties of the particles. Crystallographers often explicitly state the type of illumination used when referring to a method, as with the terms X-ray diffraction, neutron diffraction, and electron diffraction.

Three Types of Crystallography

The three types of radiation—X-rays, electrons, and neutrons—interact with the specimen in different ways. X-rays interact with the spatial distribution of the valence electrons, while electrons are charged particles and therefore feel the total charge distribution of both the atomic nuclei and the surrounding electrons. Neutrons are scattered by the atomic nuclei through the strong nuclear forces, but in addition, the magnetic moment of neutrons is non-zero. They are therefore also scattered by magnetic fields. Because of these different forms of interaction, the three types of radiation are suitable for different crystallographic studies.

Theory

An image of a small object is usually generated by using a lens to focus the illuminating radiation, as is done with the rays of the visible spectrum in light microscopy. However, the wavelength of visible light (about 4000 to 7000 Angstroms) is three orders of magnitude longer then the length of typical atomic bonds and atoms themselves (about 1 to 2 Angstroms). Therefore, obtaining information about the spatial arrangement of atoms requires the use of radiation with shorter wavelengths, such as X-rays. Employing shorter wavelengths implied abandoning microscopy and true imaging, however, because there exists no material from which a lens capable of focusing this type of radiation can be created. (That said, scientists have had some success focusing X-rays with microscopic Fresnel zone plates made from gold, and by critical-angle reflection inside long tapered capillaries). Diffracted x-ray beams cannot be focussed to produce images, so the sample structure must be reconstructed from the diffraction pattern. Sharp features in the diffraction pattern arise from periodic, repeating structure in the sample, which are often very strong due to coherent reflection of many photons from many regularly spaced instances of similar structure, while non-periodic components of the structure result in diffuse (and usually weak) diffraction features.

Because of their highly ordered and repetitive structure, crystals give diffraction patterns of sharp Bragg reflection spots, and are ideal for analyzing the structure of solids.

Technique

Some materials studied using crystallography, proteins for example, do not occur naturally as crystals. Typically, such molecules are placed in solution and allowed to crystallize over days, weeks, or months through vapour diffusion. A drop of solution containing the molecule, buffer, and precipitants is sealed in a container with a reservoir containing a hygroscopic solution. Water in the drop diffuses to the reservoir, slowly increasing the concentration and allowing a crystal to form. If the concentration were to rise more quickly, the molecule would simply precipitate out of solution, resulting in disorderly granules rather than an orderly and hence usable crystal.

Once a crystal is obtained, data can be collected using a beam of radiation. Although many universities that engage in crystallographic research have their own X-ray producing equipment, synchrotrons are often used as X-ray sources, because of the purer and more complete patterns such sources can generate. Synchrotron sources also have a much higher intensity of X-ray beams, so data collection takes a fraction of the time normally necessary at weaker sources.

Producing an image from a diffraction pattern requires sophisticated mathematics and often an iterative process of modeling and refinement. In this process, the mathematically predicted diffraction patterns of a hypothesized or "model" structure are compared to the actual pattern generated by the crystalline sample. Ideally, researchers make several initial guesses, which through refinement all converge on the same answer. Models are refined until their predicted patterns match to as great a degree as can be achieved without radical revision of the model. This is a painstaking process, made much easier today by computers.

The mathematical methods for the analysis of diffraction data only apply to *patterns,* which in turn result only when waves diffract from orderly arrays. Hence crystallography applies for the most part only to crystals, or to molecules which can be coaxed to crystallize for the sake of measurement. In spite of this, a certain amount of molecular information can be deduced from the patterns that are generated by fibers and powders, which while not as perfect as a solid crystal, may exhibit a degree of order. This level of order can be sufficient to deduce the structure of simple molecules, or to determine the coarse features of more complicated molecules (the double-helical structure of DNA, for example, was deduced from an X-ray diffraction pattern that had been generated by a fibrous sample).

Materials Engineering

Crystallography is a tool that is often employed by materials scientists. In single crystals, the effect of the crystalline arrangement of atoms is often easy to see macroscopically, because the natural shapes of crystals reflect the atomic structure. In addition, physical properties are often controlled by crystalline defects. The understanding of crystal structures is an important prerequisite for understanding crystallographic defects. Mostly, materials do not occur in a single crystalline, but poly-crystalline form, such that the powder diffraction method plays a most important role in structural determination.

A number of other physical properties are linked to crystallography. For example, the minerals in clay form small, flat, plate like structures. Clay can be easily deformed because the plate like particles can slip along each other in the plane of the plates, yet remain strongly connected in the direction perpendicular to the plates. Such mechanisms can be studied by crystallographic texture measurements.

In another example, iron transforms from a body-centered cubic (bcc) structure to a face-centered cubic (fcc) structure called austenite when it is heated. The fcc structure is a close-packed structure, and the bcc structure is not, which explains why the volume of the iron decreases when this transformation occurs.

Crystallography is useful in phase identification. When performing any process on a material, it may be desired to find out what compounds and what phases are present in the material. Each phase has a characteristic arrangement of atoms. Techniques like X-ray diffraction can be used to identify which patterns are present in the material, and thus which compounds are present note that the determination of the "phases" within a material should not be confused with the more general problem of "phase determination," which refers to the phase of waves as they diffract from planes within a crystal, and which is a necessary step in the interpretation of complicated diffraction patterns.

Crystallography covers the enumeration of the symmetry patterns which can be formed by atoms in a crystal and for this reason has a relation to group theory and geometry. See symmetry group.

Biology

X-ray crystallography is the primary method for determining the molecular conformations of biological macromolecules, particularly protein and nucleic acids such as DNA and RNA. In fact, the double-helical structure of DNA was deduced from crystallographic data. The first crystal structure of a macromolecule, myoglobin, was solved in 1958.

The Protein Data Bank (PDB) is a freely accessible repository for the structures of proteins and other biological macromolecules. Computer programs like RasMol or Pymol can be used to visualize biological molecular structures.

Electron crystallography has been used to determine some protein structures, most notably membrane proteins and viral capsids.

References

- Ceramic-engineering: newworldencyclopedia.org, Retrieved 10 March, 2019
- Bartosz Tylkowski; Karolina Wieszczycka; Renata Jastrząb, eds. (2017-09-25). Polymer Engineering. ISBN 9783110469745. OCLC 1011405606
- Processes-of-metallurgy, chemistry: byjus.com, Retrieved 14 January, 2019
- Metallurgy, science: britannica.com, Retrieved 5 April, 2019

- Meister, John J. Polymer modification : principles, techniques, and applications. ISBN 9781482269819. OCLC 1075130719
- Nanotechnology, technology: britannica.com, Retrieved 17 May, 2019
- Nanomaterials: understandingnano.com, Retrieved 9 February, 2019
- "Nonlinear Viscoelasticity", Polymer Engineering Science and Viscoelasticity (PDF), Springer US, 2008, pp. 327–364, doi:10.1007/978-0-387-73861-1-10, ISBN 9780387738604
- Crystallography: newworldencyclopedia.org, Retrieved 4 June, 2019

Permissions

All chapters in this book are published with permission under the Creative Commons Attribution Share Alike License or equivalent. Every chapter published in this book has been scrutinized by our experts. Their significance has been extensively debated. The topics covered herein carry significant information for a comprehensive understanding. They may even be implemented as practical applications or may be referred to as a beginning point for further studies.

We would like to thank the editorial team for lending their expertise to make the book truly unique. They have played a crucial role in the development of this book. Without their invaluable contributions this book wouldn't have been possible. They have made vital efforts to compile up to date information on the varied aspects of this subject to make this book a valuable addition to the collection of many professionals and students.

This book was conceptualized with the vision of imparting up-to-date and integrated information in this field. To ensure the same, a matchless editorial board was set up. Every individual on the board went through rigorous rounds of assessment to prove their worth. After which they invested a large part of their time researching and compiling the most relevant data for our readers.

The editorial board has been involved in producing this book since its inception. They have spent rigorous hours researching and exploring the diverse topics which have resulted in the successful publishing of this book. They have passed on their knowledge of decades through this book. To expedite this challenging task, the publisher supported the team at every step. A small team of assistant editors was also appointed to further simplify the editing procedure and attain best results for the readers.

Apart from the editorial board, the designing team has also invested a significant amount of their time in understanding the subject and creating the most relevant covers. They scrutinized every image to scout for the most suitable representation of the subject and create an appropriate cover for the book.

The publishing team has been an ardent support to the editorial, designing and production team. Their endless efforts to recruit the best for this project, has resulted in the accomplishment of this book. They are a veteran in the field of academics and their pool of knowledge is as vast as their experience in printing. Their expertise and guidance has proved useful at every step. Their uncompromising quality standards have made this book an exceptional effort. Their encouragement from time to time has been an inspiration for everyone.

The publisher and the editorial board hope that this book will prove to be a valuable piece of knowledge for students, practitioners and scholars across the globe.

Index

www.ingramcontent.com/pod-product-compliance
Lightning Source LLC
LaVergne TN
LVHW080653210926
848485LV00003B/372